SECURITIES AND EXCHANGE COMMISSION: PROGRAMS AND OPERATIONS

SECURITIES AND EXCHANGE COMMISSION: PROGRAMS AND OPERATIONS

UNITED STATES GOVERNMENT ACCOUNTABILITY OFFICE

Nova Science Publishers, Inc.

New York

For permission to use material from this book please contact us:
Telephone 631-231-7269; Fax 631-231-8175
Web Site: http://www.novapublishers.com

NOTICE TO THE READER

The Publisher has taken reasonable care in the preparation of this book, but makes no expressed or implied warranty of any kind and assumes no responsibility for any errors or omissions. No liability is assumed for incidental or consequential damages in connection with or arising out of information contained in this book. The Publisher shall not be liable for any special, consequential, or exemplary damages resulting, in whole or in part, from the readers' use of, or reliance upon, this material.

Independent verification should be sought for any data, advice or recommendations contained in this book. In addition, no responsibility is assumed by the publisher for any injury and/or damage to persons or property arising from any methods, products, instructions, ideas or otherwise contained in this publication.

This publication is designed to provide accurate and authoritative information with regard to the subject matter covered herein. It is sold with the clear understanding that the Publisher is not engaged in rendering legal or any other professional services. If legal or any other expert assistance is required, the services of a competent person should be sought. FROM A DECLARATION OF PARTICIPANTS JOINTLY ADOPTED BY A COMMITTEE OF THE AMERICAN BAR ASSOCIATION AND A COMMITTEE OF PUBLISHERS.

LIBRARY OF CONGRESS CATALOGING-IN-PUBLICATION DATA

United States. Government Accountability Office.
 Securities and Exchange Commission : programs and operations / GAO.
    p. cm.
 ISBN 978-1-60692-045-9 (softcover)
 1. United States. Securities and Exchange Commission. 2. Securities--United States. 3. Stock exchanges--United States. I. Title.
 HG4910.U544 2008
 332.64'273--dc22
                        2008035978

Published by Nova Science Publishers, Inc. ✛ New York

CONTENTS

PREFACE

The Securities and Exchange Commission's (SEC) Division of Enforcement (Enforcement) plays a key role in meeting the agency's responsibility to enforce securities laws and regulations. While Enforcement has brought a number of high-profile cases, questions have been raised over how effectively the division manages its operations and resources. This book: (1) discusses the structure of the SEC's inspection program for SROs; (2) evaluates certain aspects of the SEC's inspection program; (3) describes the SRO referral process and evaluates the SEC's information system for receiving SRO referrals; (4) outlines GAO's evaluation of enforcements investigation planning and information systems and (5) oversight of the Fair Fund program; (6) describes OCIE's revisions after 2003 to the examination approach for investment companies and investment advisers; (7) discusses OCIE's compliance with its examination exit procedures; and (8) describes reforms OCIE implemented since January 2006 to enhance, among other things, communication with registrants.

Chapter 1

OPPORTUNITIES EXIST TO IMPROVE OVERSIGHT OF SELF-REGULATORY ORGANIZATIONS*

ABBREVIATIONS

ALJ	administrative law judge
ARP	Automation Review Policy
CATS	Case Activity Tracking System
FINRA	Financial Industry Regulatory Authority
IG	Inspectors General
IT	information technology
MUI	matter under investigation
NYSE	New York Stock Exchange
OCC	Office of the Comptroller of the Currency
OCIE	Office of Compliance Inspections and Examinations
OIT	Office of Information Technology
OMS	Office of Market Surveillance
ORSA	Options Regulatory Surveillance Authority
SEC	Securities and Exchange Commission
SRO	self-regulatory organization
UAF	unusual activity file

---

* Excerpted from CRS Report GAO-08-33, dated November 2007.

November 15, 2007

The Honorable
Charles E. Grassley
Ranking Member
Committee on Finance United States Senate

Dear Senator Grassley:

Self-regulatory organizations (SRO) include, among others, national securities exchanges and securities associations registered with the Securities and Exchange Commission (SEC), such as the New York Stock Exchange (NYSE) and the Financial Industry Regulatory Authority (FINRA).[1] At the time that the system of self-regulation was created, Congress, regulators, and market participants recognized that this structure possessed inherent conflicts of interest because of the dual role of SROs as both market operators and regulators. Nevertheless, Congress adopted self-regulation, as opposed to direct federal regulation of the securities markets, to prevent excessive government involvement in market operations, which could hinder competition and market innovation. Also, Congress concluded that self-regulation with federal oversight would be more efficient and less costly to taxpayers.

For industry self-regulation to function effectively, SEC must ensure that SROs are fulfilling their regulatory responsibilities. As regulators, SROs are primarily responsible for establishing the standards under which their members conduct business; monitoring the way that business is conducted; bringing disciplinary actions against their members for violating applicable federal statutes, SEC's rules, and their own rules; and referring potential violations of nonmembers to SEC's Division of Enforcement (Enforcement). SEC oversees SROs through such actions as reviewing their rule proposals and information technology (IT) security through its Division of Market Regulation (Market Regulation), and periodically inspecting their operations through its Office of Compliance Inspections and Examinations (OCIE). OCIE inspections are intended to assess the effectiveness of SRO operations and often make recommendations intended to improve them.[2] If OCIE finds that an SRO has failed to comply with, or enforce member compliance with, SRO rules or federal securities laws, it may refer the SRO to Enforcement for further investigation and potential sanctions. More recently, recognizing the role of internal controls in promoting compliance and effectiveness within SROs, OCIE has begun focusing increased attention on the activity and work products of the internal audit function at SROs.

This chapter addresses your interest in the actions taken by SEC to ensure that SROs—in particular, the two largest SROs, NASD (the SRO that provided market oversight of the NASDAQ Stock Market and certain other exchanges prior to FINRA) and NYSE—are fulfilling their regulatory responsibilities by effectively monitoring and investigating suspicious trading in listed securities and, where appropriate, prosecuting misconduct involving member broker-dealers or referring potential misconduct by non-SRO members to SEC.[3] It also addresses your interest in SEC's processes for managing and acting upon referrals received by Enforcement from SROs. Specifically, this report

1. discusses the overall structure of SEC's inspection program and, more specifically, its approach to inspections of SRO surveillance, investigative, and disciplinary programs (enforcement programs);
2. evaluates certain aspects of SEC's inspection program, including guidance and planning, the use of SRO internal audit products, and the tracking of inspection recommendations; and
3. describes the SRO referral process to SEC's Enforcement Division and recent trends in referral numbers and related SEC investigations, and evaluates SEC's information system for advisories and referrals.

To address our first objective, we reviewed and analyzed OCIE documentation of the 11 inspections completed between March 2002 and January 2007 of NASD and NYSE enforcement programs, an OCIE memorandum to the Commission describing the SRO inspection process, and our prior work. Furthermore, we observed a demonstration of various IT systems that NASD used to monitor the markets and track investigations and disciplinary actions. We also conducted interviews with staff from OCIE, NASD, and NYSE. To address our second objective, we reviewed OCIE inspection guidance related to the review of SRO internal audit reports, guidance for bank examiners from the Board of Governors of the Federal Reserve System (Federal Reserve) and the Office of the Comptroller of the Currency (OCC), inspection guidelines developed by the Inspectors General (IG), and our prior work. In addition, we reviewed SRO internal and external audits of IT security and interviewed staff from OCIE, Market Regulation, NASD, and NYSE. Furthermore, we reviewed internal control standards for the federal government and conducted interviews with OCIE and Enforcement officials on their respective procedures for ensuring that SROs implement inspection recommendations and remedial actions required as part of enforcement actions. In addition, we reviewed and summarized the enforcement actions brought by SEC against SROs between 1995 and 2007. To address our third objective, we observed a demonstration from Enforcement staff on the division's system for receiving SRO referrals, and we interviewed Enforcement, NASD, and NYSE staff to determine how SEC manages SRO referrals and conducts investigations. To understand trends in SRO referrals and SEC investigations related to these referrals, we requested and analyzed data from SEC's referral receipt system and case tracking system from fiscal years 2003 through 2006. We inquired about checks that SEC performs on these data and determined they were reliable for our purposes.

We performed our work in Washington, D.C.; New York, New York; and Rockville, Maryland, between September 2006 and September 2007 in accordance with generally accepted government auditing standards. Appendix I provides a more detailed description of our scope and methodology.

RESULTS IN BRIEF

To help ensure that SROs are fulfilling their regulatory responsibilities, OCIE conducts both routine and special inspections of SRO regulatory programs. Routine inspections assess SRO enforcement, arbitration, listings, and member examination programs at regular intervals. Special inspections are conducted as warranted and encompass follow-up work on prior recommendations or enforcement actions, investigations of tips or reports, and sweep inspections.[4] OCIE's process for conducting SRO inspections includes performing background research, drafting a planning memorandum, conducting on-site reviews, holding exit interviews, and drafting a written inspection report that is reviewed and approved by the Commission. Inspection teams consist of a lead attorney and from 2 to 6 other staff reporting to an OCIE branch chief. The number of staff dedicated to SRO inspections has fluctuated in recent years, increasing from 36 to 62 between fiscal years 2002 and 2005 in response to an increase in SEC funding, but then subsequently decreasing over the following 2 years to 46 as of June 2007. OCIE officials attributed this decline to staff attrition and a recent SEC-wide hiring freeze. OCIE officials told us that inspections of SRO enforcement programs are intended to assess the design and operation of the programs to determine whether they effectively fulfill regulatory responsibilities. In these inspections, OCIE assesses the parameters of SRO surveillance systems, reviews the adequacy of SRO policies and procedures, and reviews SRO case files to determine whether SRO staff handled the resulting alerts and investigations in compliance with its policies and procedures. OCIE inspections may result in recommendations that are intended to address any deficiencies identified and improve the effectiveness of SROs.

While OCIE inspections have assessed and made recommendations to improve the effectiveness of SRO enforcement programs, we identified several opportunities for OCIE and Market Regulation to enhance their oversight of SROs by developing formal guidance, leveraging the work of SRO internal audit functions, and enhancing information systems. The following points summarize our key findings on SEC's inspection program:

- Although examiners have processes for inspecting SRO enforcement programs, OCIE has not documented these processes in an examination manual or other formal guidance. According to OCIE officials, the uniqueness of SRO rules and surveillance systems would make it difficult to tailor a manual to all SROs and keep it current. However, other federal financial regulators that perform inspections of diverse and complex organizations have developed guidelines or standards that outline the objectives of the inspection program and functional approaches to meeting the objectives, and inspection standards developed by the IG community recommend developing and implementing written policies and procedures for internal controls over inspection processes to provide reasonable assurance over conformance with an organization's policies and procedures. Similar documentation by OCIE could help ensure uniform standards and quality controls and serve as a reference guide for new examiners.

- OCIE officials said that they focus their inspection resources on areas judged highest risk by considering factors such as when an area was last inspected, the size of the program, the results of past inspections and consultations with other SEC offices and divisions. However, OCIE's risk-assessment and inspection planning processes do not incorporate information gathered by SRO internal audit functions. Our previous work has shown that SRO internal audits covered aspects of their regulatory programs that OCIE also inspected, and could be useful for OCIE's planning purposes. In contrast, risk assessments of large banks that federal bank examiners conduct are partly based on internal audit reports, and examiners may adjust their plans to avoid duplication of effort and minimize burden to the banks. By not considering internal audit information in their risk-assessment and planning processes, OCIE examiners may be duplicating SRO efforts or missing opportunities to direct examination resources to other higher-risk or less-examined program areas.[5]

- Market Regulation could also enhance SEC oversight over SROs by further leveraging information from SRO internal audit functions regarding the security of their enforcement-related databases. These databases contain critical information about the disciplinary and other regulatory history of SRO members; SEC and other regulators rely on the accuracy and integrity of these data for conducting their own investigative and enforcement efforts. While Market Regulation staff conduct regular security reviews of IT systems that SEC and SROs consider important to trading operations, in accordance with SEC guidance, as well as those systems used to remit regulatory fees, these reviews are not intended to directly address the security of enforcement-related systems. NASD and NYSE internal and external auditors regularly review the security of these systems, and have generally concluded that these SROs have adequate controls in place. However, because Market Regulation does not review these reports, it has little knowledge about the comprehensiveness of SRO reviews and cannot determine whether SROs have taken the appropriate steps to secure enforcement-related information or what risks a security breach could pose.

- OCIE currently does not formally track the implementation status of inspection recommendations. Rather, according to OCIE, management consults with staff to obtain such information as needed. The number of recommendations in 11 inspection reports we reviewed ranged from 4 to 29, although OCIE officials said some inspections resulted in as many as 30 or 40 recommendations. Without formal tracking, OCIE's ability to efficiently and effectively generate and evaluate trend information—such as patterns in the types of deficiencies found or the implementation status of recommendations across SROs, or over time—as well as to develop performance measures on the effectiveness of its inspection program, may be limited. OCIE officials told us that OCIE is currently working with SEC's Office of Information Technology (OIT) to develop a new tracking system and software that will allow OCIE to generate management reports from this system in 2008.

SRO advisories—information on suspicious trading activity that does not rise to the level of a referral—and referrals, which are received electronically in Enforcement, have increased in recent years, as have related SEC investigations and enforcement actions, but the information system SEC uses to receive advisories and referrals has limitations. SROs send advisories and referrals electronically to Enforcement's Office of Market Surveillance (OMS). The advisories and referrals, which may lead to an investigation and enforcement actions, undergo multiple reviews. OMS staff apply general criteria, such as the nature of the entity and the alleged market activity, to determine whether advisories and referrals merit further review and investigation by Enforcement attorneys. Our review of SEC data found that advisories from SROs grew significantly from 5 in fiscal year 2003 to 190 in fiscal year 2006. During the same period, referrals from SROs grew from 438 to 514, or an increase of 17 percent. Numbers of SEC investigations and enforcement actions also showed a corresponding increase. We found that almost 91 percent of all advisories and almost 80 percent of SRO referrals sent to SEC during this period involved suspected insider trading activity, which Enforcement and SRO staff attribute to increased merger and acquisition activity. Although SEC received and processed an increasing number of advisories and referrals during the review period, its systems for receiving them and tracking the resulting investigations have limited capabilities for searching and analyzing information. For example, the SRO referral system only allows users to search advisories or referrals by the issuer whose stock was flagged by SRO surveillance, not by the names of individuals or hedge funds that may be associated with the suspicious trading activity. Furthermore, the referral and case tracking systems are not linked and do not allow staff to readily analyze advisory and referral trends or characteristics, such as the duration of SRO and SEC processes for receiving and responding to SRO referrals. Combined, these limitations may reduce the ability of Enforcement staff to manage the advisory and referral processes by efficiently accessing information that could help monitor unusual market activity and make decisions about opening investigations.

This report makes three recommendations designed to strengthen SEC's oversight of SROs. In summary, we recommend that the SEC Chairman (1) establish a written framework for conducting inspections of SRO enforcement programs, and broaden current guidance to SRO inspection staff to have them consider to what extent they will use SRO internal audit reports when planning SRO inspections; (2) ensure that Market Regulation makes certain that SROs include in their periodic risk assessment of their IT systems a review of the security of their enforcement-related databases, and that Market Regulation reviews the comprehensiveness and completeness of the related SRO-sponsored audits of SRO enforcement-related databases; and (3) ensure that any software developed for tracking SRO inspections includes the ability to track SRO inspection recommendations, and consider IT improvements that would increase staff's ability to search for, monitor, and analyze information on SRO advisories and referrals.

We provided a draft of this report to SEC, and the agency provided written comments that are reprinted in appendix V. In its written comments, SEC agreed with our recommendations. In response to our recommendations, SEC said that OCIE will provide its SRO inspectors with written guidance with respect to its risk-scoping techniques and compiled summary of inspection practices; will assess the quality and reliability of SRO internal audit programs and determine whether, and the extent to which, inspections can be risk-focused on the basis of SRO internal audit work; and is developing a database for, among other things, tracking the implementation status of SRO inspection recommendations. Furthermore, Market Regulation will implement our recommendation to ensure that enforcement-related databases continue to be periodically reviewed by SRO internal audit programs and that these reviews are comprehensive and complete, and Enforcement plans to consider the recommended system improvements. SEC also provided technical comments on the draft report, which were incorporated in this report as appropriate.

BACKGROUND

SROs are responsible for the surveillance of the trading activity on their markets.[6] Market transactions take place on electronic or floor-based platforms. SROs employ electronic surveillance systems to monitor market participants' compliance with SRO rules and federal securities laws. Electronic surveillance systems are programmed to review trading and other data for aberrational trading patterns or scenarios within defined parameters. Also, SROs review trading as a result of complaints from the public, members, and member firms and as a result of required notifications, such as those concerning offerings. One of the key surveillance systems employed by SROs monitors the markets for insider trading. We discuss SRO surveillance systems and investigatory procedures related to insider trading in more detail in appendix II.[7]

SRO staff review alerts generated by the electronic surveillance systems to identify those that warrant further investigation. When SROs find evidence of potential violations of securities laws or SRO rules involving their members, they can conduct disciplinary hearings and impose penalties.

These penalties can range from disciplinary letters to the imposition of monetary fines to expulsion from trading and SRO membership. SROs do not have jurisdiction over entities and individuals that are not part of their membership, and, as such, any suspected violations on the part of nonmembers are referred directly to Enforcement. SROs maintain records of their investigations and the resulting disciplinary actions as part of their internal case tracking systems. In addition, as part of their market surveillance efforts, SROs, such as NASD and NYSE, maintain databases with information on individuals and firms associated with suspicious trading activity, such as insider trading. NASD also maintains the Central Registration Depository, the securities industry online registration and licensing database. This database makes complaint and disciplinary information about registered brokers and securities firms available to the public and, in more detailed form, to SEC, other securities regulators, and law enforcement authorities.

OCIE administers SEC's nationwide examination and inspection program. Within OCIE, the Office of Market Oversight primarily focuses on issues related to securities trading activities, with the objective of evaluating whether SRO enforcement programs and procedures are adequate for providing surveillance of the markets, investigating potential violations, and disciplining violators under SRO jurisdiction. OCIE also inspects other SRO regulatory programs, which include, among others, arbitration, listings, sales practice, and financial and operational programs. As part of the latter, OCIE coordinates the compliance inspections of NASD's district offices, which are responsible for examining broker-dealer members for compliance with SRO rules and federal securities laws.

In cases where OCIE discovers potentially egregious violations of federal securities laws or SRO rules during an SRO inspection, it may refer the case to Enforcement, which is responsible for further investigating these potential violations; recommending Commission action when appropriate, either in a federal court or before an administrative law judge (ALJ); and negotiating settlements.

SEC's Market Regulation administers and executes the agency's programs relating to the structure and operation of the securities markets, which include regulation of SROs and review of their proposed rule changes. SEC has delegated authority to Market Regulation for other aspects of SRO rulemaking as well, including the authority to publish notices of proposed rule changes and to approve proposed rule changes.

OCIE APPROACH TO SRO INSPECTIONS FOCUSES ON DETERMINING WHETHER SROS IDENTIFY VIOLATIONS AND ENFORCE AND COMPLY WITH SRO RULES EFFECTIVELY

OCIE conducts both routine and special inspections of SRO regulatory programs as part of its oversight efforts. We found that the SRO inspection process generally includes a planning phase, an on-site review of SRO programs, and a written report to the SRO documenting inspection findings and recommendations that is reviewed and approved by the Commission. OCIE typically staffs inspections with a lead attorney and from 2 to 6 other staff, who also work concurrently on at least 1 other SRO inspection. The number of staff dedicated to SRO inspections has fluctuated in recent years, but as of September 2007 totaled 46. According to OCIE officials, inspections of SRO enforcement programs are intended to assess the design and operation of SRO enforcement programs to determine if they effectively fulfill SRO regulatory responsibilities. As part of these inspections, OCIE takes steps to assess SRO surveillance systems, reviews SRO policies and procedures for investigating potential violations and disciplining violators of rules and laws, and reviews samples of SRO case files to determine whether SRO staff were complying with the policies and procedures.

Overall Structure of OCIE Program Encompasses Routine Inspections of Key Regulatory Programs at SROs as Well as Special Inspections

As part of its SRO oversight responsibilities, OCIE conducts both routine and special inspections of SRO regulatory programs. At regular intervals, OCIE conducts routine inspections of key regulatory programs, such as SRO enforcement, arbitration, examination, and listings programs.[8] The inspection cycles are based on the size of the SRO market and the type of regulatory program, with key programs of larger SROs, such as NYSE and NASD, being inspected from every 1 to 2 years, and smaller regional SROs from every 3 to 4 years.[9] Inspection of enforcement programs typically include a review of SRO surveillance programs for identifying potential violations of trading rules or laws, investigating those potential violations, and disciplining those who violate the rule or law. While sometimes OCIE conducts a comprehensive review of these programs, especially at the smaller SROs, often these inspections focus on a specific aspect of the programs, such as fixed income. We discuss OCIE's process for targeting their routine inspections later in this report. OCIE also conducts special inspections of SRO regulatory programs, as warranted. Special inspections typically originate from a tip or need to follow up on past inspection findings and recommendations. Special inspections also can include sweep inspections, where OCIE probes specific activities of all SROs or a sample of them to identify emerging compliance issues. According to OCIE officials, some aspect of every SRO is generally examined every year through a routine examination of a specific regulatory program or through a special inspection.

OCIE's inspection process for SROs generally includes a planning phase, an on-site review and analysis, and a final inspection report to the SRO (see figure 1). During inspection planning, OCIE identifies the SRO program to be inspected and assigns staff who conduct initial research on the program, prepare materials for each individual inspection on the basis of the inspection's focus, and draft a planning memorandum. In preparation for the on-site inspection, OCIE typically sends an initial document request to the SRO, asking for general program information such as organizational charts and copies of SRO policies and procedures or, if OCIE is reviewing a surveillance program, logs of alerts and the resulting investigations. We discuss OCIE's review of enforcement programs in more detail later in this section. After reviewing the documents provided, staff plan the on-site phase of the inspection, which can include additional requests for specific documents, such as case files, to be made available for review while on-site. OCIE staff typically spends 1 week on-site interviewing SRO staff and reviewing SRO case files and other documentation. After the on-site visit, OCIE staff continue their analysis in the home office; conduct follow-up interviews or request additional documentation, as needed; and begin drafting the inspection report. Staff present their initial inspection findings and recommendations to the SRO in an exit interview and incorporate initial SRO responses into the draft inspection report. Once the report is drafted, staff then circulate it to other interested SEC divisions and offices—such as the Office of General Counsel, Market Regulation, or Enforcement—for their review and comment, and then submit the report to the Commission for review. Following Commission consideration and authorization, staff issue a nonpublic report to the SRO and request that the SRO respond in writing within a specified time frame, typically 30 days.[10]

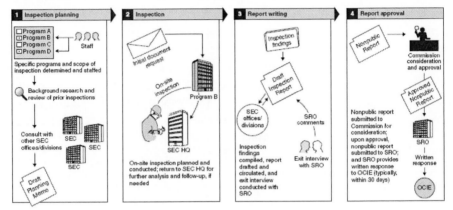

Sources: GAO (data); Art Explosion (images).

Figure 1. Key Steps in OCIE's Inspection Process for SROs.

According to OCIE officials, they staff SRO inspections with a lead attorney and from 2 to 6 other staff reporting to an OCIE branch chief. These individuals are typically staffed concurrently on at least 1 other SRO inspection. As shown in table 1, as of September 2007, the SRO inspection group consisted of 46 staff, including 14 managers, 29 examiners, and 3 other support staff. Of the 32 examiners and support staff, 16 are dedicated to market oversight inspections.[11]

Table 1. Number of OCIE Staff Delegated to SRO Inspections, Fiscal Years 2002-2007 (September)

Fiscal year	Managers			Staff		Year total
	Senior officer	Assistant director	Branch chief	Professional	Support	
2002	2	2	4	25	3	36
2003	2	2	6	27	3	40
2004	2	3	9	41	4	59
2005	2	4	9	43	4	62
2006	2	3	9	29	4	47
2007 (through Sept. 2007)	2	4	8	29	3	46

Source: OCIE.

Table 1 shows that between fiscal years 2002 and 2005, SRO inspection staffing increased from 36 to 62, or 72 percent. OCIE staff said that this increase was largely due to the increase in funding SEC received as a result of the Sarbanes-Oxley Act of 2002.[12] Since then, SRO inspection staffing has declined from 62 to 46, or 26 percent, which OCIE officials attributed to staff attrition and the inability of OCIE to hire replacements during a SEC-wide hiring freeze that occurred from May 2005 to October 2006. OCIE officials stated that despite the decrease in staff numbers, they have continued to conduct routine inspections on schedule, although the inspections may last longer than usual. Also, they said that they have not been able to do as many special inspections as they otherwise would have conducted. OCIE officials told us that the SRO inspection group recently received 6 additional professional staff positions, which it is now in the process of filling.[13]

OCIE Assesses Design and Operation of SRO Enforcement Programs to Determine Whether SROs Effectively Fulfill Their Regulatory Responsibilities

According to OCIE officials, inspections of SRO enforcement programs are intended to assess the design and operation of SRO enforcement programs to determine whether they effectively identify violations, enforce compliance among members, and follow their own procedures. More specifically, OCIE officials said that when inspecting SRO surveillance programs, their objectives are to determine whether (1) the parameters of SRO electronic surveillance systems are appropriately designed to generate exceptions that identify potential instances of noncompliance with SRO rules and federal securities laws and (2) the systems are effectively detecting such activity. When reviewing SRO surveillance systems, OCIE begins by asking the SRO for copies of the exchange rules that it is required to enforce, a description of the coding behind the surveillance systems designed to monitor the markets for compliance with these rules, and logs of the alerts that these systems generated. OCIE staff then review this information to determine whether the system is appropriately designed to identify noncompliance and whether it is functioning as designed. For example, as part of one inspection, OCIE staff found that the parameters of a specific surveillance system were too restrictive, after observing that the system did not generate any alerts over the inspection period. Conversely, OCIE staff said that if in reviewing a surveillance system, the inspection team saw that the system generated 10,000 alerts every quarter, they would follow up with the SRO to determine whether the indications of numerous rule violations were plausible or whether the parameters of the system were set appropriately. Either way, they said that the inspection team would dedicate resources to looking at that system.

Similarly, when evaluating SRO programs for investigating potential violations of SRO rules or federal securities laws and disciplining broker-dealer members, OCIE officials stated that their objective is to determine whether (1) SRO policies and procedures are appropriately designed to uncover violations of SRO rules and federal securities laws and to administer the appropriate disciplinary measures and (2) the SRO is complying with these policies and procedures. OCIE staff first request copies of the relevant policies and procedures for investigating surveillance alerts and for disciplining members found to be in violation of SRO rules and federal securities laws. They also ask for lists of the resulting investigations and enforcement actions, including referrals on nonmembers to SEC. OCIE staff then analyze this information to assess the extent to which SRO policies and procedures direct the SRO staff to conduct thorough reviews and investigations and, when applicable, to take appropriate disciplinary action. For example, during a recently completed sweep inspection of SRO surveillance and investigative programs related to insider trading, OCIE evaluated related SRO policies and procedures for reviewing alerts and opening investigations to determine whether they directed staff to coordinate appropriately with other SROs. We discuss the results of this sweep inspection—including a plan that the options SROs submitted to SEC to create a more uniform and coordinated method for the regulation, surveillance, investigation, and detection of insider trading—in appendix II. As part of another inspection we reviewed, OCIE found that an SRO had not yet developed formal procedures for its analysts to review alerts that were generated by a recently implemented surveillance system. OCIE recommended that the SRO develop such procedures.

When reviewing SRO enforcement programs, OCIE also assesses whether the SRO is in compliance with its own policies and procedures. To accomplish this objective, OCIE staff select and review case files pertaining to a sample of alerts, investigations, and disciplinary files from the lists that they have asked the SRO to generate. OCIE staff said when reviewing these files, they pay particular attention to the strength of the evidence upon which the SRO analyst relied in determining whether to close an alert or an investigation or to refer the case to SRO enforcement, SEC, or other appropriate regulators. In this way, OCIE staff said they can evaluate whether the SRO is enforcing its rules and federal securities laws consistently among its members and, in the case of certain federal laws such as those prohibiting insider-trading, between members and nonmembers. For example, in one inspection we reviewed, OCIE found that the

SRO used its informal disciplinary measures inappropriately when disciplining its members, and recommended that formal disciplinary actions be taken when informal actions had already occurred.

OCIE inspections may result in recommendations to SROs that are intended to address any deficiencies identified and to improve SRO effectiveness. OCIE officials said that for SRO enforcement programs, they tend to make recommendations flexible enough to allow SROs to implement them in a manner that best fits their unique business models and surveillance systems. As we have previously discussed, if OCIE finds serious deficiencies at an SRO, it can refer the case to Enforcement. Such referrals are relatively infrequent—between January 1995 and September 2007, SEC brought and settled 10 enforcement actions against SROs (see app. III). According to OCIE officials, recommendation follow-up is primarily the responsibility of the examination team, under the supervision of the assistant director assigned to the inspection. Inspection follow-up begins with evaluating written responses by SROs to the inspection report and obtaining documentation of SRO efforts to address the recommendations, and can continue for several years, depending on the complexity of the recommendation. For example, OCIE officials said that some recommendations, such as those that involve the design and implementation of new information technology, may require continued dialogue with the SRO over several years before the recommendation is fully implemented. OCIE also may follow up on inspection recommendations during a subsequent inspection of the SRO. OCIE officials said that in the event the SRO does not take steps to address a recommendation that staff believe is critical, they can elevate the matter to OCIE management or the Commission, although they said that this happens infrequently. We discuss the tracking of inspection recommendations later in this report.

WRITTEN INSPECTION GUIDANCE, INCREASED LEVERAGING OF SRO INTERNAL AUDIT PRODUCTS, AND IT IMPROVEMENTS COULD ENHANCE SEC OVERSIGHT OF SROS

We identified several opportunities for OCIE and Market Regulation to enhance their oversight of SROs by developing formal guidance, leveraging the work of SRO internal audit functions, and enhancing information systems. First, although OCIE has developed a general process for inspecting SRO enforcement programs, it has not developed an examination manual or other formal guidance for examiners to use when conducting inspections, as it has for examinations of other market participants. Such guidance could help OCIE ensure that its inspection procedures and products are subject to uniform standards and quality controls. Second, OCIE has recently expanded the use of the SRO internal and external audit reports while on-site at the SRO; however, OCIE does not leverage this work in the planning process, which could result in duplication of effort and missed opportunities to better target inspection resources. Third, in accordance with SEC policy, Market Regulation regularly inspects SRO IT systems related to market operations for adequate security controls and reviews related to SRO internal audit reports. However, this review does not target SRO enforcement-related databases, which contain investigative and disciplinary information that SROs maintain and upon which other regulators rely. Finally, OCIE currently does not formally track the implementation status of inspection recommendations, which ranged as high as 29 in the inspections that we reviewed. The lack of formal tracking may reduce OCIE's ability to efficiently and effectively generate and evaluate trend information, such as patterns in the types of deficiencies found or the implementation status of recommendations across SROs, or over time.

Lack of Formal Guidance for Inspections of SRO Enforcement Programs Could Limit OCIE's Ability to Ensure Staff Compliance with Internal Controls

Our interviews with OCIE officials and reviews of selected inspection workpapers indicated that OCIE examiners typically follow a general process when conducting reviews of SRO enforcement programs. This process begins with examination planning, is followed by data gathering, and ends with reporting. However, OCIE has not developed an examination manual or other formal guidance for its examiners to use when conducting inspections of SRO enforcement programs. According to OCIE officials, because SRO rules and corresponding surveillance systems are unique and constantly evolving, it would be difficult to develop a detailed inspection manual that could be tailored to all SROs and also remain current. These officials said that an examination manual is not necessary to ensure consistency among SRO inspections because the SRO inspection group is a relatively small group within OCIE, and all of the staff are centralized in headquarters. On the other hand, they said that because OCIE's inspection program for investment companies, investment advisers, and broker-dealers has hundreds of examiners across SEC headquarters and its regional offices who are responsible for examining thousands of firms, OCIE has developed detailed inspection manuals to ensure consistency across examinations of these firms. Similarly, OCIE officials said that they have developed guidelines for SRO examiners conducting oversight inspections of NASD's district offices because OCIE relies on examination staff in the SEC regional offices to assist them in conducting these inspections.

In contrast to OCIE, federal banking regulators, such as the Federal Reserve and OCC, have developed written guidance for the examination of large banks—also highly complex and diverse institutions—that outlines the objectives of the program and describes the processes and functional approaches used to meet those objectives. By not establishing written guidance for conducting inspections of SRO enforcement and other regulatory programs, OCIE may be limiting its ability to ensure that its inspection processes and products are subject to basic quality controls in such areas as examination planning, data collection, and report review. For example, in several of the inspections we reviewed, we did not find evidence of supervisory review, which is a key aspect of inspection quality control. According to OCIE officials, the team leader is expected to review the work of team members. However, without written policies and procedures specifying how and when this review is to be conducted and documented, it is difficult to establish whether the team leaders comply with this quality control. According to inspection standards developed by the IG community, each organization that conducts inspections should develop and implement written policies and procedures for internal controls over its inspection processes to provide reasonable assurance over conformance with organizational policies and procedures. As another example, when conducting inspections of SRO enforcement programs, OCIE officials said that team leaders often require their teams to use data collection instruments, such as checklists, when reviewing SRO files to ensure a consistent and complete review of all of the files selected, particularly when there are inexperienced staff on the team. While potentially an effective means of collecting data, according to OCIE officials, the decision to use these tools is up to the individual team leader, and not all teams employ them. According to IG inspection standards, evidence developed under an effective system of internal controls generally is more reliable than evidence obtained where such controls are lacking. By not establishing standards addressing quality controls in data collection, OCIE's ability to ensure the consistency and reliability of data collected across its SRO inspection teams may be limited. Furthermore, without written guidelines, new examiners lack a reference tool that could facilitate their orientation in the inspection program.

OCIE's Limited Use of SRO Internal Audit Reports in Inspection Planning May Diminish Opportunities to Better Target Inspection Resources

While OCIE employs a risk-based approach to conducting SRO inspections, OCIE's risk-assessment and inspection planning processes do not incorporate information gathered through SRO internal audits. According to OCIE officials, OCIE tailors inspections of SRO programs (particularly at the two largest SROs) to focus on those areas judged to pose the greatest risk to the SRO or the general market. In determining which areas present the highest risk, OCIE officials said they consider such factors as the amount of time that has passed since a particular area was last inspected, the size of the area, the results of past inspections, and consultations with other SEC offices and divisions. For example, because the enforcement programs at NASD and NYSE encompass hundreds of surveillance systems, OCIE officials said examiners cannot review all systems as part of one inspection. As a result, OCIE officials said examiners first conduct a preliminary analysis of requested documents and focus inspection resources on those systems or areas that are judged to pose the greatest risk. According to OCIE officials, because the regional SROs have smaller programs, OCIE staff typically are able to conduct a more comprehensive review of the entire enforcement program during a single inspection.

We previously recommended that OCIE develop and implement a policy requiring examiners to routinely use SRO internal review reports in planning and conducting SRO inspections.[14] Prior to October 2006, OCIE's practice was to request SRO internal audit reports only when OCIE believed specific problems existed at an SRO. In October 2006, OCIE issued a memorandum broadening the circumstances in which OCIE would request and use these reports. The memorandum directs examiners to request that SROs make all internal audit reports related to the program area under inspection available for the staff's on-site review, including workpapers or any reviews conducted by any regulatory quality review unit of the SRO or an outside auditor. According to the memorandum, on-site review of these reports may be useful in determining whether the SRO has identified particular areas of concern in a program area and adequately addressed those problems, assessing whether an SRO addressed prior inspection findings and recommendations, and helping staff determine whether they should limit or expand their review of particular issues during an inspection.

OCIE staff said that in fiscal year 2008, they also plan to begin reviewing the internal audit functions of SROs, with the goal of determining whether SRO internal audit functions are effective. For example, OCIE officials said that they plan to evaluate whether the internal audit functions are independent of SRO management, conduct thorough reviews of all relevant areas (particularly, regulatory programs), and have sufficient staffing levels. OCIE officials said that as part of their reviews, they also plan to assess the quality and reliability of SRO internal audit reports and assess whether SROs have implemented the recommendations resulting from these reports. OCIE officials told us that they are in the planning phase of this review, and, as such, they have not yet developed written guidance for their examiners in conducting these reviews.[15]

While OCIE's October 2006 memorandum broadened the use of SRO internal audit reports to encompass on-site reviews during inspections, it did not address the use of internal audit reports for planning purposes, as we had recommended. In contrast, the risk assessments of large banks that federal bank examiners conduct during the planning phase are based, in part, on internal audit reports, and examiners may adjust their examination plans to avoid duplication of effort and minimize burden to the bank. For example, according to examination guidance that the Federal Reserve issued, to avoid duplication of effort and burden to the institution, examiners may consider using these workpapers and conclusions to the extent that examiners test the work performed by the internal or external auditors and determine it is reliable. Similarly, examination guidance issued by OCC states that examiners' assessments of a bank's audit and control functions help leverage OCC resources, establish the scope of current and future supervisory activities, and assess the quality of risk management.

By not considering the work and work products of SRO internal audit functions in its inspection planning process, OCIE examiners may be duplicating SRO efforts, causing regulatory burden, or missing opportunities to direct examination resources to other higher-risk or less-examined program areas. For example, our previous work, which focused on the listing programs of SROs, showed that SRO internal audit functions had examined or were in the process of examining aspects of their listing programs that OCIE had covered in its most recent inspections, and that resulting reports could be useful to OCIE in planning as well as conducting inspections.[16] As OCIE begins to assess the quality of SRO internal audit functions and work products, the opportunity exists for OCIE to further leverage these products in targeting its own inspection efforts. OCIE officials said that as part of their upcoming reviews of SRO internal audit functions, they will assess whether SRO internal audit products may be helpful in assisting them in targeting inspections of particular SRO functions.

OCIE could also further leverage the work performed by SRO internal and external auditors to monitor a particular regulatory program between inspections. In our review of OCIE inspections of NASD and NYSE enforcement programs, as many as 8 years passed between inspections of a particular surveillance system and related investigations and disciplinary actions. Moreover, as OCIE officials noted, the recent decline in SRO inspection staff has lengthened the time it takes to complete a routine SRO inspection and limited their ability to conduct additional special inspections. Unless OCIE regularly informed itself of the results of SRO efforts to review these systems, it may not know of emerging or resurgent issues until the next inspection.[17]

SEC Does Not Obtain Information on the Security of SRO Enforcement-Related Systems and Databases

As we have previously discussed, SROs conduct surveillance of trading activity on their markets; carry out investigations; and bring disciplinary proceedings involving their own members or, when appropriate, make referrals to SEC when the suspicious activity involves nonmembers. However, SEC's Market Regulation does not obtain information on the security of SRO enforcement-related databases—IT applications for storing data about SRO investigations and disciplinary actions taken against SRO members—when conducting reviews of IT security at SROs. Under SEC's Automation Review Policy (ARP), Market Regulation conducts on-site reviews of SRO trading systems, information dissemination systems, clearance and settlement systems, and electronic communications networks and makes recommendations for improvements when necessary.[18] Market Regulation also conducts reviews of SRO general and application controls over the collection of fees under section 31 of the Securities Exchange Act of 1934.[19] These are IT systems designated for remitting fees to SEC as part of the section 31 program, which ensures that the data produced by these systems are authorized, and completely and accurately processed and reported.

Market Regulation officials said that they do not target enforcement-related databases for specific review, since the ARP policy statement is specifically intended to oversee systems essential to market operations. These officials said that Market Regulation could include a review of the security of enforcement-related databases both in their general assessments of SRO IT infrastructure security within the ARP and in section 31 reviews. They explained that both of these reviews include testing of components and evaluations of general access controls and changes made within SRO organizationwide network structures in their routine reviews of specific IT programs and systems, such as SRO computer operations, security assessments, internal and external audit IT coverage, and systems outage notification procedures and systems change notifications. However, these general assessments by Market Regulation would not necessarily provide SEC with information on potential risks specific to the security of the data contained in enforcement-related databases.

NASD and NYSE officials told us that they conduct their own regular internal inspections of security of IT systems, which include reviews of enforcement-related databases. In addition, both SROs contract with external companies that regularly conduct reviews of the security controls of their technology systems. We reviewed several of these internal and external audits, which include reviews of SRO enforcement-related systems and databases conducted from fiscal years 2002 through 2006. These reviews generally concluded that NASD and NYSE have adequate controls in place to protect sensitive enforcement-related data.

The internal and external audit reports of NYSE and NASD that we reviewed showed that these reports could be a valuable source of information for Market Regulation on specific risks to enforcement-related databases. Market Regulation officials said that in conducting ARP-related inspections, they review SRO internal and external audit reports related to the infrastructure of SRO IT systems; however, they do not specifically look for information related to the assessment of security of enforcement-related databases. In addition, SEC staff said that although they generally receive all the internal and external audit reports done of SRO systems relating to trading and clearing functions, they may not always receive such reports relating to other systems, including enforcement-related databases, from all SROs.

Since SROs, SEC, and other regulators rely on the accuracy and integrity of the data in SRO enforcement-related databases in fulfilling their own regulatory responsibilities, protecting this information from unauthorized access is critical to regulatory efforts. For example, as we discuss later in this report, SEC uses SRO surveillance data in carrying out its own enforcement efforts related to securities trading. Furthermore, SROs are responsible for maintaining complaint and disciplinary data on their members—information that is essential for identifying recidivists. By not periodically obtaining information to ensure that the SRO risk-assessment process and SRO-sponsored audits continue to be included in SRO assessment cycles and that the audits are comprehensive and complete, Market Regulation cannot assess whether SROs have taken the appropriate steps to ensure the security of sensitive enforcement-related information, or the level of risk that a data breach could pose.

Lack of Formal Tracking System May Limit OCIE's Ability to Effectively Assess SRO Implementation of Inspection Recommendations

Although OCIE officials said that they have worked with SROs to address the intent of recent inspection recommendations, we were not able to readily verify the status of the recommendations in the inspections we reviewed because OCIE does not formally track inspection recommendations or the status of their implementation. OCIE officials said that when OCIE management is interested in obtaining an update on the recommendations resulting from an inspection, they consult directly with the examination team assigned to the SRO inspection. OCIE officials also said that they do not consider the lack of a formal tracking system to have affected their ability to manage any follow-up of inspection recommendations because there are relatively few SROs, and OCIE staff is in frequent contact with them. OCIE's informal methods for tracking inspection recommendations contrast with the expectations set by federal internal control standards for ensuring that management has relevant, reliable, and timely information regarding key agency activities.[20] These standards state that key information on agency operations should be recorded and communicated to management and others within the entity and within a time frame that enables management to carry out its internal control and other responsibilities.

Without a formal tracking system, the ability of OCIE management to effectively and efficiently monitor the implementation of SRO inspection recommendations and conduct programwide analyses may be limited. Of the 11 inspections of NASD and NYSE enforcement programs we reviewed, the number of recommendations OCIE made ranged from 4 to 29, with an average of 11.[21] They also ranged in complexity, from asking the SRO to update its policies and procedures to recommending that an SRO implement an entire surveillance program. For example, we observed recommendations calling for, among other things, improving case file documentation, changing the parameters of a surveillance system, implementing an automated tracking system, and improving SRO member education. OCIE officials said that some inspections resulted in as many as 30 or 40 recommendations. Without a formal tracking system, OCIE management must rely on staff's availability and ability to recall recommendation-related information, which may be reliable when discussing an individual inspection, but may limit OCIE management's ability to efficiently generate and evaluate trend information, such as patterns in the types of deficiencies found or the implementation status of recommendations across SROs, or over time. Implementing a formal tracking system would not only allow management to more robustly assess the recommendations to SROs and their progress in implementing them, but would allow it to develop performance measures that could assist management in evaluating the effectiveness of its inspection program.

According to OCIE and SEC's OIT officials, OCIE recently began working with OIT to develop a new examination tracking system that will include the capability to track SRO responses and implementation status of OCIE recommendations. OCIE officials said that planned requirements for the system includes a field to enter the recommendation, a field for OCIE inspectors to broadly categorize the status of its implementation, and a text box for inspectors to elaborate on the recommendation and its implementation status. OCIE officials also said that they expect that the system will be able to trace the history of a recommendation. OIT officials told us that they are developing separate software that will allow OCIE to generate management reports using data from the tracking systems as well as other database; however, the requirements for any management reports OCIE would receive have yet to be determined. According to an OCIE official, the recommendation tracking system and reporting capabilities may be an effective way to provide OCIE management with a high-level characterization of implementation status. OCIE officials said that in response to our concerns, they plan to deploy an interim, stand-alone recommendation tracking system that will provide a management report, in the form of a spreadsheet, that contains all open recommendations to SROs resulting from SRO inspections and the current status of SRO efforts to implement them. These officials said that they expect to use this spreadsheet until the previously described OIT projects are implemented in 2008.

SRO ADVISORIES AND REFERRALS HAVE INCREASED, AS HAVE RELATED SEC INVESTIGATIONS AND ENFORCEMENT ACTIONS, BUT INFORMATION SYSTEMS FOR ADVISORIES AND REFERRALS HAVE LIMITATIONS

Enforcement receives advisories and referrals, which undergo multiple stages of review and may lead to opening an investigation, through an electronic system in OMS. After opening investigations, Enforcement further reviews the evidence gathered to decide whether to pursue civil or administrative actions, or both. From fiscal years 2003 to 2006, OMS received an increasing number of advisories and referrals from SROs, such as NYSE and NASD, most of which involved insider trading. However, limited search capabilities of the SRO system and the lack of a link between the SRO and case activity tracking systems have limited Enforcement staff's ability to electronically search advisory and referral information, monitor unusual market activity, make decisions about opening matters under inquiry (MUI) and investigations, and assess case activities.

OMS Uses a Multistep Process to Review SRO Referral Information that Can Lead to Opening Investigations and Subsequent Enforcement Actions

Upon receipt of SRO information in its Web-based SRO Referral Receipt System (SRO system), OMS makes initial decisions on referrals and forwards selected referral materials to investigative attorneys. After initial reviews by OMS staff, Enforcement may decide to open investigations if it determines evidence garnered during its inquiry period warrants doing so and staff and financial resources are available. If investigation evidence merits, staff may pursue administrative or civil actions and seek remedies, such as cease-and-desist orders and civil monetary penalties.

Enforcement Receives Advisories and Referrals from SROs about Unusual Market Activity through a Web-Based System

The referral process begins when OMS staff receive SRO advisories and referrals on unusual market activity through a secure Web-based electronic system called the SRO system. SEC officials noted that SRO referrals help SEC identify and respond to unusual market activity by those who are not members of SROs, investigate those suspected of potentially illegal behavior, and take action when the circumstances of cases and evidence are appropriate. OMS branch chiefs, who are responsible for reviewing advisories and referrals, access the SRO system on a weekly basis to review all SRO-submitted advisories and referrals.

SRO advisories and referrals usually consist of a short form with basic background information on the suspected unusual market activity by SRO nonmembers that includes the name of the security issuer, date of the unusual activity, and a description of the market activity identified by the SRO. The materials also contain a text attachment, which includes more detailed narrative information, such as a chronology of unusual activity and specific information about issuers and individuals potentially associated with that activity. SEC does not receive information electronically or otherwise on unusual market activity by SRO members or related investigations by SROs of the unusual member activity.

OMS Reviews Both Advisories and Referrals, and Forwards Referrals to Enforcement Attorneys for Possible Investigatory Action

After reading advisories and referrals, OMS branch chiefs use SEC's National Relationship Search Index, an electronic system that connects to and works with a range of other SEC systems, such as the Case Activity Tracking System (CATS), to determine whether existing SEC investigations involve the issuer noted in the SRO advisory or referral.[22] If an investigation already exists that involves the issuer noted in the advisory or referral, the branch chiefs will forward the advisory or referral to the Enforcement attorney conducting that investigation for review and incorporation into his or her case.

If Enforcement has not already opened an investigation on a particular issuer, OMS staff store advisories in the SRO system, but do not investigate them because they do not contain information as detailed as that found in referrals in the SRO system.[23] However, SROs may continue their market surveillance efforts on an advisory, further develop information on the unusual market activity, and submit all information later as a referral for potential action by SEC. For referrals, branch chiefs apply criteria—such as (1) the nature of the unusual market activity, (2) the persons involved and their employment positions, (3) the dollar value of the unusual activity in question, (4) potential harm to the financial markets and individual investors, and (5) any other information branch chiefs may have obtained through conversations with SRO staff—to make initial decisions about the merit of forwarding the referrals to Enforcement management and attorneys for possible SEC investigation. Enforcement associate directors review and either approve or disapprove branch chiefs' recommendations about the referrals. Referrals not recommended by branch chiefs for approval are stored in the SRO system and may be accessed as needed.

If approved, OMS branch chiefs open an MUI, a 60-day initial inquiry period, and electronically forward all referral information to SEC headquarters or the appropriate regional office, where investigative attorneys and management have up to 60 days to review all available case information and consider staff and financial resources to decide whether to proceed with a full investigation. Once the MUI has been opened, Enforcement staff assigns the MUI a CATS case number, and staff use CATS to track all components of the case until it is closed.[24] Figure 2 outlines SEC's process and average time frames for receiving, processing, and investigating unusual market activity identified by SROs.

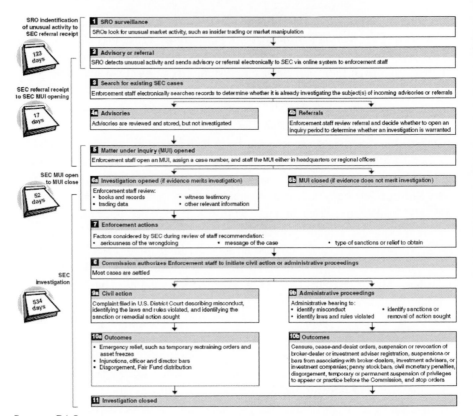

Source: GAO.

Figure 2. SEC's Process and Average Time Frames for Receiving SRO Advisories and Referrals and Conducting Related Investigations.

Enforcement staff at headquarters or the regional offices use criteria that are similar to those used by OMS staff during their initial review, but also consider the level of financial resources available for investigations and the availability of Enforcement staff to determine whether to close the MUI or open an investigation. If Enforcement staff do not open an investigation, the MUI is closed in CATS and staff document the reason(s) for closure, which may include insufficient evidence, resource limitations, or a newly opened case being merged with an existing case.

When Evidence from Investigation Merits, Enforcement Division Can Pursue Civil and Administrative Actions

If the Enforcement Division develops evidence it deems sufficient for moving forward, SEC may institute civil or administrative enforcement actions, or both. When determining how to proceed, Enforcement staff consider such factors as the seriousness of the wrongdoing, the technical nature of the matter under investigation, and the type of sanction or relief sought. When the misconduct warrants it, SEC will bring both types of proceedings. With civil actions, SEC files a complaint with a federal district court that describes the misconduct, identifies the laws and rules violated, and identifies the sanction or remedial action that is sought. For example, SEC often seeks civil monetary penalties and the return of illegal profits, known as disgorgement. The courts also may bar or suspend an individual from serving as a corporate officer or director (see figure 2).

SEC can seek a variety of sanctions through administrative enforcement proceedings as well. An ALJ, who is independent of SEC, presides over a hearing and considers the evidence presented by the Enforcement staff as well as any evidence submitted by the subject of the proceeding. Following the hearing, the ALJ issues an initial decision, which contains a recommended sanction. Administrative sanctions or outcomes include cease-and-desist orders, suspension or revocation of broker-dealer and investment adviser registration, censures, bars from association with certain persons or entities in the securities industry, payment of civil monetary penalties, and return of illegal profits. Both Enforcement staff and the defendant may appeal all or any portion of the initial decision to SEC Commissioners, who may affirm the decision of the ALJ, reverse the decision, or remand it for additional hearings. An SRO may also agree to undertake other remedial actions in a settlement agreement with SEC.

Once civil or administrative proceedings have concluded and all outcomes are finalized, SEC closes the investigation and terminates the case in CATS.[25]

Figure 2 also provides data on the durations involved with referral and investigation processes and shows that stages of the process—from SRO identification of unusual market activity to the closure of investigations—vary in their duration. We analyzed data SEC provided from its referral and case tracking systems from fiscal years 2003 to 2006. For those cases for which the data had open and close dates for the investigation stage of the process, it took an average of 726 days or almost 2 years from the point that SROs identify unusual market activity and send SEC referrals to the time that SEC completely investigates and concludes cases.[26] Of this total time, it took, on average, 192 days for the first three steps in the process, which include SROs identifying unusual market activity and referring it to SEC and SEC opening an MUI to conduct its initial inquiry on referrals.[27] It took, on average, another 534 days for SEC to investigate that unusual market activity; institute administrative or civil enforcement proceedings; administer outcomes, such as issuing and collecting fines; and completely close investigations.[28]

From Fiscal Years 2003 through 2006, the Number of SRO Advisories and Referrals and SEC Investigations and Enforcement Actions Significantly Increased

Data we reviewed from SEC's SRO system and CATS showed that the number of advisories, referrals, and investigations significantly increased from fiscal years 2003 through 2006. More specifically, advisories increased from 5 in fiscal year 2003 to 190 in fiscal year 2006 and totaled 390 for the period. Of the 4-year total, 354, or 91 percent, were insider trading advisories, and an additional 3 percent involved market manipulation issues. Data from SEC's SRO system on 1,640 referrals showed that the number of referrals SEC received from SROs grew from 438 in fiscal year 2003 to 514 in fiscal year 2006, an increase of 17 percent. Of the total number of referrals, almost 80 percent involved suspected insider trading activities. In addition, NYSE and NASD submitted 1,095, or almost 70 percent, of the total number of referrals. SEC and SRO officials attributed the increase to more merger and acquisition activity in the marketplace.

Data SEC provided to us from its case tracking system showed a corresponding increase in the number of investigations SEC opened from SRO referrals over the same period. The number of investigations rose from 82 in fiscal year 2003 to 208 in fiscal year 2006, an increase of 154 percent. Case actions, which follow SEC's determination of whether to file a case as an administrative proceeding or a civil action, also increased. The number of case actions rose from 2 in fiscal year 2003 to 29 in fiscal year 2006. SEC actions result in case outcomes such as permanent injunctions, preliminary injunctions, restraining orders, administrative proceeding orders, and emergency actions. These case outcomes rose from 3 in fiscal year 2003 to 82 in fiscal year 2006. Case outcomes also may include "relief," such as disgorgement, payment of prejudgment interest and other monetary penalties, asset freezes, and officer and director bans.

For example, in 2003, NYSE referred unusual market activity to SEC after suspecting potential insider trading activity. After opening an MUI and investigating the activity, the case resulted in an administrative proceeding and a civil action. The case resulted in a range of outcomes against 6 individuals. The administrative proceeding specifically resulted in an order barring individuals alleged in the case from associating with one another on trading. The civil action resulted in permanent injunctions to stop the suspected use of material, nonpublic information and in financial penalties that included disgorgement.

Figure 3 illustrates the upward trend in the numbers of advisories, referrals, MUIs, investigations, case actions, and case outcomes for the period we reviewed.[29] The figure also shows that more than three quarters of the referrals were made for insider trading. Market manipulation and "other" activity, including activity associated with issuer reporting and financial disclosure and initial securities offerings, constituted the other major categories of referrals. Appendix IV provides additional data on these trends by fiscal year.

Fiscal year		Number of:									
	Advisories		Referrals		MUIs		Investigations		Actions		Outcomes
2003	5		438		153		82		2		3
2004	50		340		220		136		8		23
2005	145		348		265		148		23		45
2006	190		514		322		208		29		82
Total	390		1,640		960		574		62		153

Other

Market manipulation

Insider trading

Source: GAO.

Figure 3. SRO Advisories and Referrals, and Related SEC MUIs, Investigations, Actions, and Outcomes, Fiscal Years 2003-2006.

Limited Search Capabilities of the SRO System and Lack of Linkage to Case Tracking System May Limit Management of Process and Staff Analysis

SEC's SRO system featured limited capability to electronically search information on advisories and referrals and may limit Enforcement staff's ability to efficiently monitor unusual market activity, make subsequent decisions about opening MUIs and investigations, and manage the SRO advisory and referral process. As we have previously discussed, federal internal control standards state that management needs relevant, reliable, and timely communications relating to internal and external events. In addition, these standards state that the information should be distributed in a form and time frame that permits management and others who need it to perform their duties efficiently.

SEC developed the SRO system to receive and store advisory and referral information from SROs and enable SEC staff to make initial decisions about which SRO-identified market activities to investigate. The system primarily receives information on unusual market activity based on SRO surveillance of trades among stock issuers. This information includes the name of the security issuer; the date of the unusual activity; and a description of the type of activity, among other data. The SRO system also stores narrative attachments, which the SROs provide to SEC, that contain additional information about individuals or entities, such as investment advisers or hedge funds, associated with unusual market activity. While the system allows OMS staff to search by issuer, the narrative information cannot be easily searched in the system; instead, the attachments must be individually opened and read. An Enforcement branch chief noted that narrative information can help establish patterns of behavior that are critical when SEC tries to investigate potentially fraudulent activity, such as market manipulation and insider trading. Furthermore, only OMS branch chiefs have access to the SRO system, so attorneys who need that information have to consult with OMS branch chiefs or contact SRO staff directly, rather than access that information electronically. In addition, since the referral receipt and case tracking systems are not linked, management is unable to readily assess the efficiency and effectiveness of the referral and investigation processes. For example, SEC is unable to extract information from a single source on how long it takes both SROs and SEC to work through different stages of cases over time, from referral receipt (SRO system) to opening MUIs and conducting investigations (case tracking system).[30] SEC headquarters and regional office officials noted that receiving information in a timely manner is critical to the investigative steps of assembling the facts of the case and collecting evidence on those potentially involved with unusual market activity. To obtain this information and customized reports and statistics on Enforcement operations, division officials said they must submit requests to SEC's OIT and then wait for OIT staff to respond to the request. As noted in our 2007 report on Enforcement Division operations, these requests may take several days to 1 week to complete. Having recognized system limitations, SEC officials have undertaken efforts to make improvements to CATS by developing a new case information management system called the Hub. However, these planned improvements do not address limitations of the SRO system and do not include expanded linkages between the SRO system and CATS.[31]

CONCLUSIONS

SEC's oversight of SRO enforcement programs has produced positive outcomes. For example, in response to an OCIE recommendation, SROs in the options market have developed a new surveillance authority, which is intended to improve coordination among SROs in monitoring the markets for insider trading and investigating any resulting alerts. The equities markets are expected to soon follow with a similar plan. SEC, through its Enforcement Division, has worked with SROs to detect and respond to potential securities laws violations. Between fiscal years 2003 and 2006, SEC responded to an increasing number of SRO referrals—a large percentage of which are related to insider trading—with an increasing number of investigations and enforcement actions. SEC has started to incorporate the results of SRO internal audits into its on-site inspections, which helps to leverage resources. In addition, the agency plans to expand its oversight of SRO functions to include reviews of the internal audit function—with an emphasis on independence, staffing levels, and scope of coverage. Such reviews could help ensure that SROs are effectively assessing risks, instituting appropriate controls, and carrying out their responsibilities.

However, several opportunities exist to enhance the efforts used by SEC to oversee SROs and, particularly, their enforcement programs. Specifically, OCIE examiners are conducting inspections of SRO enforcement programs without formal guidance. Although our review of a sample of inspections found that examiners have developed a methodology for reviewing SRO enforcement programs, the lack of written guidance—which establishes minimum standards and quality controls—could limit OCIE's ability to provide reasonable assurances that its inspection processes and products are subject to basic quality controls in such areas as examination planning, data collection, and report review. Moreover, the lack of formal guidance could result in individual inspection teams creating data collection and other examination tools that otherwise would be centralized and more efficiently shared across inspection teams.

Furthermore, OCIE's recent internal guidance on the use of SRO internal audit-related reports does not address the use of these reports for risk-assessment and inspection planning purposes, as we have previously recommended. We continue to believe that the use of these reports when conducting risk assessments and determining the scope of an upcoming inspection could allow OCIE to better leverage its inspection resources, especially if OCIE determines that the reports produced by SRO internal audit functions are reliable. As OCIE officials noted, they plan to begin assessing SRO internal audit functions in 2008, including the quality and reliability of their work products, although they have not yet developed guidance for inspection staff on conducting these reviews. By not considering the work and work products of the SRO internal audit function in its inspection planning process, OCIE may be duplicating SRO efforts and not maximizing the use of its limited resources. OCIE also may be missing an opportunity to better monitor the effectiveness of the SRO regulatory programs (including enforcement programs) between inspections.

SEC also has an opportunity to leverage the work of SRO internal audit functions in its assessment of information security at SROs. Since ARP Policy Statements specifically are intended to oversee systems essential to market operations, Market Regulation officials do not target enforcement-related databases for specific review. Although SROs have assessed the security controls of these databases, Market Regulation officials have little knowledge of the content or comprehensiveness of these audits. As a result, Market Regulation cannot determine whether SROs have taken the appropriate steps to ensure the security of this sensitive information. Market Regulation could facilitate this evaluation by making certain that enforcement-related databases continue to be periodically reviewed by SROs, and that these reviews are comprehensive and complete.

Both OCIE and Enforcement could benefit from improvements to information technology systems when overseeing SROs. OCIE currently lacks a system that tracks the status of inspection recommendations. OCIE officials told us that a new examination tracking database is in development that will allow OCIE to track the implementation of inspection recommendations as well as software that will allow OCIE to generate management reports from this database. By ensuring these system capabilities, OCIE management could improve its ability to monitor the implementation of OCIE recommendations, and begin developing measures for assessing the effectiveness of its program.

Finally, while SEC has responded to a significant increase in SRO referrals between fiscal years 2003 and 2006, Enforcement's systems for receiving referrals and tracking the resulting investigations have limited capabilities for searching and analyzing information related to these referrals. Enforcement is currently working to address some limitations in its case tracking system; however, this effort does not include making improvements to the separate system used to receive and manage SRO referrals. By including system improvements to allow electronic access to all of the information contained in advisories and referrals submitted by SROs, generate management reports, and provide links to the case tracking system, Enforcement could enhance its ability to efficiently and effectively manage SRO advisories and referrals and conduct analyses that could contribute to improved SEC planning, operations, and oversight.

APPENDIX I. SCOPE AND METHODOLOGY

To discuss the overall structure of the Securities and Exchange Commission's (SEC) inspection program—more specifically, its approach to inspections of self-regulatory organizations' (SRO) surveillance, investigative, and enforcement programs (enforcement programs)—we reviewed and analyzed documentation of all 11 inspections that SEC's Office of Compliance Inspections and Examinations (OCIE) completed from March 2002 through January 2007 of enforcement programs related to the former NASD and the New York Stock Exchange (NYSE). We also reviewed and analyzed an OCIE memorandum to the Commission describing the SRO inspection process, staffing data provided by OCIE, and our prior work. Furthermore, we observed a demonstration of various information technology systems that NASD used to monitor the markets and track investigations and disciplinary actions. Finally, we reviewed and summarized the enforcement actions brought by SEC against SROs from 1995 to 2007. We also conducted interviews with staff from OCIE, NASD, and NYSE.

To evaluate certain aspects of SEC's inspection program, including guidance and planning, the use of SRO internal audit products, and the tracking of inspection recommendations, we reviewed OCIE inspection guidance related to the review of NASD district offices and SRO internal audit reports, guidance for bank examiners from the Board of Governors of the Federal Reserve System and the Office of the Comptroller of the Currency, inspection guidelines developed by the inspectors general, and our prior work. In addition, we reviewed SEC guidance for conducting reviews of SRO information technology (IT) related to market trading operations and regulatory fee remittance, and NASD and NYSE internal and external audits of IT security. Furthermore, we reviewed internal control standards for the federal government and conducted interviews with officials from OCIE and SEC's Division of Enforcement (Enforcement) on their respective procedures for ensuring that SROs implement inspection recommendations and remedial actions required as part of enforcement actions. We also conducted interviews with staff from OCIE, SEC's Division of Market Regulation and Office of Information Technology, NASD, and NYSE.

To describe the SRO referral process and recent trends in referral numbers and related SEC investigations, and evaluate SEC's information system for advisories and referrals, we observed a demonstration from Enforcement staff on the capabilities of their IT systems, analyzed data from SEC's SRO Referral Receipt System (SRO system) and Case Activity Tracking System (CATS), and interviewed Enforcement, NASD, and NYSE staff to determine how SEC manages the processes for receiving SRO referrals and conducting subsequent investigations. In particular, to understand trends in SRO advisories, referrals, and subsequent SEC investigations, we requested and analyzed data from SEC's referral and case tracking systems from fiscal years 2003 through 2006. We analyzed the data to provide descriptive information on the number of SEC's advisories, referrals, matters under inquiry (MUI), investigations, actions, and case outcomes during the period. We also analyzed these data by manually merging records from the SRO system and CATS to obtain descriptive data on the amount of time it takes SROs to identify unusual market activity and convey that information to SEC, as well as how long it takes SEC to respond by opening MUIs and investigations and achieving case outcomes. We inquired about checks SEC performs on the data and deemed the data reliable for the purposes of addressing our objectives. When calculating the average duration of stages to process SRO referrals, we distinguished between case stages that featured both open and close dates and those that were open or active as of the date we received data from SEC, and we reported duration information accordingly. In addition, to calculate case stage durations, we consulted with SEC and SRO staff to distinguish between initial and updated referrals and performed duration calculations using initial referrals only to avoid double counting that could skew the average duration results.

We performed our work in Washington, D.C.; New York, New York; and Rockville, Maryland, between September 2006 and September 2007 in accordance with generally accepted government auditing standards.

APPENDIX II. SEC OVERSIGHT OF SRO ENFORCEMENT PROGRAMS RELATED TO INSIDER TRADING.

SRO surveillance, investigative, and disciplinary programs are designed to enforce SRO rules and federal securities laws related to insider trading—the buying or selling of a security by someone who has access to material, nonpublic information about the security—and are subject to SEC oversight through periodic inspections by OCIE. In January 2007, OCIE completed a sweep inspection (a probe of specific activities across all or a sample of SROs) of SRO enforcement programs related to insider trading. As a result of OCIE's inspection, the options SROs submitted a plan to SEC to create a more uniform and coordinated method for surveillance and investigation of insider trading in the options markets, and the equities SROs indicated their intent to submit a similar plan. From fiscal years 2003 through 2006, SEC significantly increased the number of investigations that related to insider trading.

SROs COORDINATE WITH SEC AND USE SURVEILLANCE, INVESTIGATIVE, AND DISCIPLINARY PROGRAMS TO ENFORCE INSIDER TRADING RULES AND LAWS

SROs employ enforcement programs to enforce SRO rules and federal securities laws related to insider trading. Insider trading is illegal because any trading that is based on this information is unfair to investors who do not have access to the information. When persons buy or sell securities on the basis of information not generally available to the public, investor confidence in market fairness can be eroded. Information that could be exploited for personal gain by insiders include such things as advance knowledge of mergers or acquisitions, development of a new drug or product, or earnings announcements. While company insiders (e.g., directors and senior executives) may be the most likely individuals to possess material, nonpublic information, others outside of the company also may gain access to the information and use it for their personal gain. For example, employees at a copy store who discovered material, nonpublic information while making presentation booklets for a firm could commit insider trading if they traded on that information prior to it being made public.

To detect insider trading, SROs have established electronic surveillance systems that monitor their markets for aberrational movements in a stock's price or volume of shares traded, among other things, and generate alerts if a stock's price or volume of shares traded moves outside of set parameters. These systems link trade activity data to news and research about corporate transactions (such as mergers, acquisitions, or earnings announcements); public databases of listed company officers and directors; and other internal and external sources of information to detect possible insider trading. For example, the NASD Securities Observation News Analysis and Regulation system combines trade activity on NASDAQ, the American Stock Exchange, and the over-the-counter markets with news stories and other external sources of information to detect potential instances of insider trading and other potential violations of federal securities laws or NASD rules.[1]

SRO staff review the thousands of alerts generated by the electronic surveillance systems annually to identify those that are most likely to involve insider trading or fraud and warrant further investigation. In conducting reviews of these alerts, SRO staff consider such factors as the materiality of news, the existence of any previous news announcements, and the profit potential. If, in reviewing the trading associated with the alert, SRO staff determines there is a strong likelihood of insider trading, they can expand this review to a full investigation. In the course of a full investigation, SROs gather information from their member broker-dealers and the issuer of the traded stock to determine whether there is any relationship between those individuals who traded the stock and those individuals who had advance knowledge of the transaction or event. For example, SRO staff will typically request from their member broker-dealers the names of individuals and organizations that traded in advance of a corporate transaction or event, a process known as bluesheeting.[2] These data are then cross-referenced with information the SRO staff obtain from the issuer of the stock, including a chronology of the events leading up to the corporate transaction or event and the names of individuals who had knowledge of inside information.

SROs have created technology-based tools to assist in the identification of potential repeat offenders. For example, SROs can compare their blue sheets to a database called the Unusual Activity File (UAF), which includes data on suspicious trading activity identified by all SROs that are part of the Intermarket Surveillance Group, to help identify persons or entities that have been flagged in prior referrals or cases related to insider trading, fraud, or market manipulation.[3] Some SROs have also developed other databases for their internal use. For example, NASD developed a database similar to the UAF for suspicious trading activity it has identified. NYSE also has developed a database of individuals who are affiliated with entities that it considers at high risk for insider trading.

When SROs find evidence of insider trading involving their members, they can conduct disciplinary hearings and impose penalties ranging from disciplinary letters to fines to expulsion from trading and SRO membership. Because SROs do not have jurisdiction over entities and individuals that are not part of their membership, they refer suspicious trading on the part of nonmembers directly to Enforcement. Although Enforcement staff do not have direct access to SRO surveillance data or recidivist databases like the UAF, several staff told us they are able to obtain any needed information from the SRO analysts who made the referrals.

Data we reviewed from NASD and NYSE between fiscal years 2003 and 2006 showed that the SROs referred significantly more nonmembers to SEC for suspected insider trading than they referred members internally to their own Enforcement staff. According to SRO staff, this may be because the majority of the entities and individuals who trade on the basis of material, nonpublic information do so as a result of connections to the issuers of the stocks traded, rather than the investment advisor role that would involve member firms and their employees. Another possible explanation, according to SRO staff, is that the individual registered persons (SRO members) typically conceal their misconduct by trading in nominee accounts or secretly sharing in the profits generated by nonregistered persons involved in the scheme. As a result, they said that concealed member misconduct is often exposed through evidence developed by SEC using its broader jurisdictional tools after the SRO has referred a nonmember to SEC. For example, they said that SEC can expose the concealed member misconduct by fully investigating the nonregistered person's activities through documents such as telephone and bank records obtained by subpoena. SEC also has the ability to issue subpoenas to nonmembers to appear for investigative testimony.

SEC's INSPECTION PROGRAM TO OVERSEE SRO ENFORCEMENT EFFORTS HAS IDENTIFIED OPPORTUNITIES FOR SROs TO IMPROVE SURVEILLANCE OF INSIDER TRADING

OCIE assesses the effectiveness of SRO regulatory programs, including enforcement programs, through periodic inspections. OCIE officials said that when evaluating SRO enforcement programs related to insider trading, their objective is to assess whether the parameters of the surveillance systems are appropriately set to detect abnormal movements in a stocks' price or volume and generate an alert, the extent to which SRO policies and procedures direct the SRO staff to conduct thorough reviews of alerts and resulting investigations, and the extent to which SRO analysts comply with these policies and procedures and apply them consistently. OCIE staff said that when reviewing case files, one of their priorities is to assess the evidence upon which the SRO analyst relied when deciding to terminate the review of an alert or investigation. For example, they said that they will assess whether the analyst selected an appropriate period to review trading records (because suspicious trades may have occurred several days or weeks prior to the material news announcement), whether the analyst reviewed the UAF and internal databases for evidence of recidivism, and whether the analyst appropriately reviewed any other stocks or entities related to the trading alert.

OCIE officials said that in light of the recent increase in merger and acquisition activity and the increased potential for insider trading, SROs are making greater efforts to detect attempts of individuals or firms to benefit on both sides of a merger or acquisition.[4] For example, they said that where previously it was common for one SRO analyst to investigate any alerts generated from the movement of the target firm and for a different analyst to investigate any alerts generated from the movement of the acquiring firm—making it difficult to identify an account or individual that may have traded on both sides of the acquisition—SRO policies now generally require one analyst to review and investigate both stocks involved in a merger or acquisition. Generally speaking, mergers and acquisitions present opportunities for insider trading because the acquiring company generally must pay more per share than the current price, causing the target firm's stock price to increase. In this case, an individual with knowledge of an upcoming acquisition could purchase the target's stock prior to the announcement and then sell for a gain the stock after the announcement at the higher price. An individual also could sell any holdings or sell short the stock of the acquiring firm if the individual believed that the acquiring firm's stock price would decrease after the announcement.[5] Finally, an individual could attempt to buy the target firm and sell (or short sell) the acquiring firm in an attempt to benefit on both sides of an acquisition.

In January 2007, OCIE completed sweep inspections of surveillance and investigatory programs related to insider trading at 10 SROs. As a result of its inspections, OCIE identified opportunities for improved coordination and standardization among SROs in monitoring and investigating possible insider trading. OCIE found that because each SRO at the time maintained its own surveillance systems, the variances in the system parameters could result in the possibility that stock or option movements might generate an alert at one SRO but not another. Furthermore, OCIE found that because each SRO was responsible for monitoring every stock that traded on its market, the SROs were duplicating the initial screening of alerts.

As a result of OCIE's then ongoing inspection, the options SROs submitted a plan to SEC to create a more uniform and coordinated method for the regulation, surveillance, investigation, and detection of insider trading in the options markets. SEC approved the plan, called Options Regulatory Surveillance Authority (ORSA), in June 2006.[6] The plan allows the options SROs to delegate part or all of the responsibility of conducting insider trading surveillance and investigations for all options trades to one or more SROs, with individual SROs remaining responsible for the regulation of their respective markets and retaining responsibility to bring disciplinary proceedings as appropriate. ORSA has currently delegated this surveillance and investigative responsibility to the Chicago Board Options Exchange. The ORSA plan also provides for the establishment of a policy committee that is responsible for overseeing the operation of the plan and for making all relevant policy decisions, including reviewing and approving surveillance standards and other parameters to be used by the SRO performing the surveillance and investigative functions under the plan. The committee also will establish guidelines for generating, reviewing, and closing insider trading alerts; specific and detailed instructions on how analysts should review alerts; and instructions on closing procedures, including proper documentation and rationale for closing an alert. OCIE officials stated that they have met regularly with the options SROs to monitor the implementation of the plan and the development of related policies and procedures. According to the Commission, the ORSA plan should allow the options exchanges to more efficiently implement surveillance programs for the detection of insider trading, while eliminating redundant effort. As a result, OCIE officials believe the plan will promote more effective regulation and surveillance.

According to OCIE officials, the equities SROs are currently drafting a similar plan for coordinating insider trading surveillance in equities markets. However, instead of designating one SRO to conduct all insider trading-related surveillance, OCIE officials said that the current draft proposal would require each listing market, or its designee, to conduct insider trading surveillance for its listed issues, regardless of where trading in the security occurred. This includes reviewing alerts, pursuing investigations, and resolving cases through referrals (to SEC) or disciplinary action. OCIE officials said that the equities SROs anticipate voting on a proposed plan at the October 2007 Intermarket Surveillance Group meeting and to submit the plan to SEC by the end of 2007.

APPENDIX III. SEC CIVIL ENFORCEMENT ACTIONS AGAINST SROS, JANUARY 1995–SEPTEMBER 2007

Pursuant to sections 19 and 21 of the Securities Exchange Act of 1934, SEC may bring enforcement actions against an SRO either in federal court or through an administrative proceeding if it has found that an SRO has violated or is unable to comply with the provisions of the act and related rules and regulations, or if it has failed to enforce member compliance with SRO rules without reasonable justification or excuse. The act authorizes SEC to seek a variety of sanctions in an administrative proceeding, including the revocation of SRO registration, issuance a cease-and-desist order, or censure. An SRO may also agree to undertake other remedial actions in a settlement agreement with SEC. In addition to the remedies available in administrative enforcement action, a district court in a civil enforcement action may impose civil monetary penalties and has discretion to fashion such other equitable remedy it deems appropriate under the circumstances.

Tables 2 through 11 summarize the 10 civil enforcement actions SEC brought against SROs from January 1995 through September 2007. For this report, we have included only those findings and terms of settlement related to SRO surveillance, investigative, or disciplinary programs (enforcement programs). As such, these summaries do not necessarily identify all findings and terms of the settlement agreements.

Table 2. Summary of Findings, Enforcement Actions, and Outcomes Brought under the SEC Administrative Proceeding of August 8, 1996

Type of action	Order Instituting Public Administrative Proceedings Pursuant to Section 19(h)(1) of the Securities Exchange Act of 1934, Making Findings and Imposing Remedial Sanctions Administrative Proceeding File No. 3-9056
Respondent	National Association of Securities Dealers (NASD)
Action date	August 8, 1996
Key findings	SEC made the following findings: • NASD failed to conduct an appropriate inquiry into an anticompetitive pricing convention among NASDAQ market makers; • NASDAQ market makers followed and enforced a pricing convention used to determine the increments in which they would adjust their displayed quotes; • market makers shared proprietary information about customer orders, collaborated and coordinated their activities, failed to honor quotations, and failed to timely report trades; and • market-making firms held excessive amounts of influence in NASD oversight, its committees, and the disciplinary process.
Outcomes	Without admitting or denying SEC's findings, NASD agreed to take the following actions: • take significant steps to restructure its governance and regulatory structure, including ensuring a substantial independent review staff reporting directly to NASDAQ's Board of Governors; • increase staff positions for Enforcement, Examination, and Market Regulation; • institute the participation of professional hearing officers to preside over disciplinary proceedings; • institute measures to enhance the enforcement of the trade reporting, firm quote, customer limit order handling, and other market-making rules; • develop an enhanced audit trail system; and • enhance its systems for trading and market surveillance.

Source: SEC.

Table 3. Summary of Findings, Enforcement Actions, and Outcomes Brought under the SEC Administrative Proceeding of June 29, 1999

Type of action	Order Instituting Public Administrative Proceedings Pursuant to Section 19(h)(1) of the Securities Exchange Act of 1934, Making Findings and Ordering Compliance with Undertakings Administrative Proceeding File No. 3-9925
Respondent	New York Stock Exchange, Inc. (NYSE)
Action date	June 29, 1999
Key findings	SEC made the following findings: NYSE • failed to enforce compliance with Section 11(a) of the Exchange Act; Rule 11a-1; and NYSE Rules 90, 95, and 111, which are aimed at preventing independent floor brokers (IFB) from exploiting their position for personal gain; • failed to take appropriate action to police the manners in which IFBs were compensated; • failed to establish surveillance procedures designed to evaluate how commissions were computed; and • suspended its routine IFB surveillance for extensive periods.
Outcomes	Without admitting or denying SEC's findings, prior to settlement with SEC, NYSE took certain steps that included: • providing new or additional guidance regarding IFB compensation arrangements; • designing and implementing a program to require the examination of all IFBs within 2-year cycles; • amending NYSE rules to require certain members to make and keep written records of compensation arrangements; • adopting new rules requiring all members to disclose their own account or accounts over which they exercise any discretion; • maintaining error accounts to facilitate NYSE monitoring for trading abuses; and beginning to develop a floor audit trail for the electronic capture of certain order information. NYSE also agreed to further take the following actions: • enhance and improve its regulation of IFBs, member firm floor brokers, specialists, registered competitive market makers, and competitive traders; • file an affidavit with the Commission setting forth the details of NYSE's compliance with the undertakings described; • retain an independent consultant for review of NYSE's rules, practices, and procedures applicable to floor members and recommend changes to these rules as necessary; and • maintain a substantial independent internal review staff with adequate resources to regularly review all aspects of NYSE.

Source: SEC.

Table 4. Summary of Findings, Enforcement Actions, and Outcomes Brought under the SEC Administrative Proceeding of September 11, 2000

Type of action	Order Instituting Public Administrative Proceedings Pursuant to Section 19(h)(1) of the Securities Exchange Act of 1934, Making Findings and Imposing Remedial Sanctions Administrative Proceeding File No. 3-10282
Respondents	American Stock Exchange (AMEX), Chicago Board Options Exchange (CBOE), Pacific Exchange (PCX), and Philadelphia Stock Exchange (PHLX)
Action date	September 11, 2000
Key findings	SEC made the following findings: • The options exchanges significantly impaired the operations of the options market by following a course of conduct under which they refrained from joint listing a large number of options. • The exchanges inadequately surveilled their markets for potential rules violations, failed to conduct thorough investigations, and failed to adequately enforce rules applicable to members on their floors. • The exchanges failed to enforce compliance with rules that promote competition, enhance investor protections, and prohibit anticompetitive conduct. • The exchanges generally lacked automated surveillance systems, and relied too heavily on complaints. • In many cases, the exchanges did not take appropriate enforcement actions when violations were uncovered. • In cases where enforcement actions were taken, the exchanges did not impose sanctions adequate to provide reasonable deterrence against future violations.
Outcomes	Without admitting or denying SEC's findings, The SROs agreed to take the following actions: • eliminate advance notice to any other market of the intention to list an existing option or new option; • eliminate any provisions to the Joint Plan that would prevent a market from commencing to list or trade any option listed on another market or an option that another market has expressed and intent to list; • enhance and improve its surveillance, investigative, and enforcement processes and activities with a view toward preventing and eliminating harassment, intimidation, refusals to deal, and retaliation against market participants acting competitively; • acting jointly, design and implement a consolidated options audit trail system; and • enhance and improve its surveillance, investigative, and enforcement processes and activities for options order handling rules, limit order displays, priority rules, trade reporting, and firm quote rules.

Source: SEC.

Table 5. Summary of Findings, Enforcement Actions, and Outcomes Brought under the SEC Administrative Proceeding of September 30, 2003

Type of action	Order Instituting Public Administrative Proceedings Pursuant to Sections 19(h) and 21C of the Securities Exchange Act of 1934, Making Findings, and Imposing a Censure, a Cease-and-Desist Order and Other Relief Administrative Proceeding File No. 3-11282
Respondent	Chicago Stock Exchange (CHX)
Action date	September 30, 2003
Key findings	SEC made the following findings: CHX • failed to implement surveillance systems and procedures to detect and prevent violations of its firm quote, trading ahead, and limit order display rules; • relied on an ineffective manual review process; • did not provide staff with adequate and consistent standards and guidelines to assist them; • failed to take adequate disciplinary action against members when violations were detected; and • failed to take adequate disciplinary action against recidivists or violators of multiple rules.
Outcomes	Without admitting or denying SEC's findings, CHX agreed to take the following actions: • increase its staffing for enforcement programs and implement new protocols and guidelines regarding surveillance; • begin offering training sessions regarding compliance with trading rules; • enhance its exception reports and computer logic; • create a regulatory oversight committee; and • hire an outside consultant to conduct a comprehensive review of CHX's trading floor surveillance and enforcement programs as well as report on its findings.

Source: SEC.

Table 6. Summary of Findings, Enforcement Actions, and Outcomes Brought under the SEC Administrative Proceeding of February 9, 2005

Type of action	Report of Investigation Pursuant to Section 21(a) of the Securities Exchange Act of 1934 Regarding the Nasdaq Stock Market, Inc., as Overseen by Its Parent, the National Association of Securities Dealers
Respondent	National Association of Securities Dealers (NASD)
Report release date	February 9, 2005
Key findings	SEC made the following findings: • NASD and NASDAQ did not adequately address a large number of wash trades and matched orders in March 2002 by MarketXT, an ECN, NASD member, and registered broker-dealer, which were reported through NASDAQ. • NASDAQ failed to communicate to NASD Regulation the observations of NASDAQ staff members relating to the trading described above. • NASDAQ supervisors failed to take any steps to ensure that the suspicious trades were referred to NASD Regulation. • NASD Regulation's automated surveillance programs did not independently detect the suspicious conduct.
Outcomes	Remedial steps taken by NASDAQ: • created a NASDAQ Regulation Group; • had the NASDAQ Office of General Counsel (OGC) formalize the procedure for responding to information that suggests a possible rule violation; • instituted mandatory companywide employee education on regulatory responsibilities; • amended its code of conduct to require that employees refer potential regulatory violations to OGC or other appropriate NASDAQ department; and • refunded the consolidated tape for the fees it received associated with MarketXT trading. Remedial steps taken by NASD: • formed a committee of the NASD board to review a number of governance issues, and studied the standards for NASD review of NASDAQ board items; • retained a law firm to review the interactions between NASD and NASDAQ in the regulatory area; and • NASD board appointed a special committee with the charge of reviewing the relationship between NASD and NASDAQ, and NASD's oversight of that relationship.

Source: SEC.

Table 7. Summary of Findings, Enforcement Actions, and Outcomes Brought under the SEC Administrative Proceeding of April 12, 2005

Type of action	Order Instituting Public Administrative Proceedings Pursuant to Sections 19(h)(1) and 21C of the Securities Exchange Act of 1934, Making Findings, Ordering Compliance with Undertakings, and Imposing a Censure and a Cease-and-Desist Order Administrative Proceeding File No. 3-11892
Respondent	New York Stock Exchange, Inc. (NYSE)
Action date	April 12, 2005
Key findings	SEC made the following findings: NYSE • failed to properly detect, investigate, and discipline widespread unlawful proprietary trading by specialists on the floor of the exchange; • surveillance systems failed to detect the vast majority of improper trades due to NYSE's reliance on automated systems whose parameters and procedures were unnecessarily and unreasonably broad; • Office of Market Surveillance policies improperly limited the cases selected for further examination; • inadequate referral procedures and investigation policies further limited the cases examined; and • additional and repeat violations were often treated with additional informal actions, rather than being escalated to formal disciplinary actions.
Outcomes	Without admitting or denying SEC's findings, NYSE agreed to take the following actions: • commit to biannual, third-party audits of its regulatory function, of which SEC receives a copy, and • establish a pilot program for sufficient audio and video equipment to capture floor trading activity occurring at a specialist's post.

Source: SEC.

Table 8. Summary of Findings, Enforcement Actions, and Outcomes Brought under the SEC Administrative Proceeding of May 19, 2005

Type of action	Order Instituting Administrative and Cease-and-Desist Proceedings Pursuant to Sections 19(h) and 21C of the Securities Exchange Act of 1934, Making Findings, and Imposing Sanctions Administrative Proceeding File No. 3-11931
Respondents	National Stock Exchange (NSX) and the CEO of NSX
Action date	May 19, 2005
Key findings	SEC made the following findings: NSX • failed to enforce compliance by its dealer firms with the market order exposure rule and the customer priority (trading ahead) rule; • did not update its interpretation after decimalization and did not bring to SEC's attention its intention to enforce the rule according to its old interpretation; • did not conduct surveillance until 2004 for violations of its customer priority rule, which prohibited designated dealers from trading ahead of customer orders in their possession; • failed to develop and implement an automated surveillance report to detect trading ahead; • when trading-ahead violations were identified, failed to perform a follow-up review of that member's trading to determine whether additional violations had occurred; and • failed to preserve e-mails made or received in the course of its business or self-regulatory activity for a minimum of 5 years.
Outcomes	Without admitting or denying SEC's findings, NSX agreed to take the following actions: • create a regulatory oversight committee (ROC); • adopt structural protections to ensure the NSX's regulatory functions shall be independent from the commercial interests of NSX and its members; • adopt internal procedures that provide for the ROC and NSX Board to approve the issuance of regulatory circulars; • create and maintain complete and detailed minutes of all NSX board meetings; • implement and maintain automated daily surveillance for potential violations of the NSX and Exchange Act rules; • require NSX designated dealers to implement system enhancements; • design and implement a mandatory training program for NSX's regulatory department that addresses compliance with the federal securities laws and NSX rules; and • hire an independent consultant to conduct a comprehensive review of NSX's policies and procedures for rulemaking, surveillance, and examination programs.

Source: SEC.

Table 9. Summary of Findings, Enforcement Actions, and Outcomes Brought under the SEC Administrative Proceeding of June 1, 2006

Type of action	Order Instituting Administrative and Cease-and-Desist Proceedings, Making Findings, and Imposing Remedial Sanctions and a Cease-and-Desist Order Pursuant to Sections 19(h) and 21C of the Securities Exchange Act of 1934 Administrative Proceeding File No. 3-12315
Respondent	Philadelphia Stock Exchange (PHLX)
Action date	June 1, 2006
Key findings	SEC made the following findings: PHLX • did not adequately surveil for violations of rules relating to priority of options orders; • failed to properly surveil for firm quote rule violations; • did not implement any type of surveillance of its equities market to monitor its specialists for compliance with the firm quote rule; • generated exception reports using improper parameters, which excluded certain transactions that were potentially priority rule or firm quote violations; • generated an excessive number of alerts and false positives in exception reports for front-running violations, making the reports ineffective; and • did not maintain adequate written surveillance procedures for PHLX investigators reviewing the surveillance reports.
Outcomes	Without admitting or denying SEC's findings, PHLX agreed to take the following actions: • obtain outside counsel and consultants to conduct a complete review of its regulatory programs, augment the ranks of regulatory staff and management, and significantly increase its regulatory budget in an effort to enhance its regulatory program and • implement a mandatory, annual training program for all floor members and members of PHLX regulatory staff responsible for surveillance, investigation, examination, and discipline of floor members that addresses compliance with the federal securities laws and PHLX rules.

Source: SEC.

Table 10. Summary of Findings, Enforcement Actions, and Outcomes Brought under the SEC Administrative Proceeding of March 22, 2007

Type of action	Order Instituting Administrative and Cease-and-Desist Proceedings, Making Findings, and Imposing Remedial Sanctions, a Censure, and a Cease-and-Desist Order Pursuant to Sections 19(h)(1) and 21C of the Securities Exchange Act of 1934 Administrative Proceeding File No. 3-12594
Respondent	American Stock Exchange (AMEX)
Action date	March 22, 2007
Key findings	SEC made the following findings: • From 1999 through June 2004, AMEX had critical deficiencies in its surveillance, investigative, and enforcement programs for ensuring compliance with its rules as well as federal securities laws. • AMEX's continual regulatory deficiencies during this period resulted in large part from its failure to pay adequate attention to regulation, put in place an oversight structure, or dedicate sufficient resources to ensure that the exchange was meeting its regulatory obligations. • AMEX failed to surveil for, or take appropriate action relating to, evidence of violations of firm quote, customer priority, limit order display, and trade reporting rules. • Under a 2000 enforcement action, the Commission ordered AMEX to enhance and improve its regulatory programs for surveillance, investigation, and enforcement of the options order handling rules. AMEX also was required to provide Commission staff with annual affirmations detailing its progress in complying with the 2000 order. AMEX failed to comply with these obligations. • AMEX employed incorrect or deficient parameters in some of its surveillance systems.
Outcomes	Without admitting or denying SEC's findings, AMEX agreed to take the following actions: • file with the Commission a proposed rule change to identify and implement enhancements to its trading systems for equities and options reasonably designed to prevent specialists from violating AMEX's priority rules; • enhance its training program and implement mandatory annual training for all floor members; • commencing in 2007, and for each of the successive 2-year periods (6 years), retain a third-party auditor to conduct a comprehensive audit of AMEX's surveillance, examination, investigative, and disciplinary programs relating to trading applicable to all floor members; and • submit an auditor's report to its board of governors and the directors of OCIE and Market Regulation, and include the audit report in its annual report. SEC ordered that AMEX shall • develop a plan of corrective action, including dates for implementation, which they are to keep and provide to the Commission upon request.

Source: SEC.

Table 11. Summary of Findings, Enforcement Actions, and Outcomes Brought under the SEC Administrative Proceeding of September 5, 2007

Type of action	Order Instituting Administrative and Cease-and-Desist Proceedings, Making Findings, and Imposing Remedial Sanctions and a Cease-and-Desist Order Pursuant to Sections 19(h) and 21C of the Securities Exchange Act of 1934 Administrative Proceeding File No. 3-12744
Respondent	Boston Stock Exchange, Inc. (BSE) and the former President of BSE
Action date	September 5, 2007
Key findings	SEC made the following findings: • BSE failed, between 1999 and 2004, to enforce certain of its rules intended to prevent BSE broker-dealer specialist firms from trading in a way that benefited them, while disadvantaging their customers who were trying to buy and sell stock. • BSE failed to develop and implement adequate procedures for surveillance of violations of its customer priority rules. • BSE's failure to implement programming changes and to otherwise conduct effective surveillance allowed hundreds, if not thousands, of violations per day to go undetected. • Violations continued even after the Commission staff had repeatedly warned BSE of the need to improve surveillance systems. • BSE internal documents demonstrated awareness of BSE's surveillance system's flaws at all levels of the organization, and these flaws resulted in the system yielding too many exceptions to be useful in detecting priority rule violations.
Outcomes	Without admitting or denying SEC's findings, prior to settlement with SEC, BSE took certain steps that included • replacement of senior management responsible for regulatory compliance during the period in which the violations discussed herein occurred. BSE also agreed to take the following actions: • Within 90 days after the issuance of the Order, enhance its existing training programs for all members of the regulatory staff responsible for surveillance, investigation, examination, and discipline. • Retain a third-party auditor, not unacceptable to the Commission, to conduct a comprehensive audit of BSE's surveillance, examination, investigation, and disciplinary programs. • The auditor must submit an audit opinion to BSE's Board of Governors, and the following Commission officials: Director of OCIE, Director of Division of Market Regulation, and Director of the Boston Regional Office. • BSE must implement the auditor's recommendations. BSE may disagree with the recommendations and may attempt to reach an agreement with the auditor. If such agreement cannot be reached, the auditor's recommendations will be binding.

Source: SEC.

APPENDIX IV. ANALYSES OF SEC-PROVIDED DATA ON VARIOUS CASE STAGES

Tables 12 to 22 include analyses of data from fiscal years 2003 to 2006 provided by SEC from its SRO system and CATS. This appendix provides specific analyses on the number and types of advisories; referrals; matters under inquiry (MUI); investigations; case actions; and case outcomes, by fiscal year and SRO. It also describes reasons that SEC closed MUIs and provides data on average and median investigation durations, by type of investigation.

Table 12. Number and Type of Advisories, Fiscal Years 2003-2006

Fiscal year	Number of insider trading advisories	Number of market manipulation advisories	Number of all other types of advisories	Total advisories
2003	5	0	0	5
2004	48	1	1	50
2005	135	3	7	145
2006	166	7	17	190
Total	354	11	25	390

Source: GAO.

Table 13. Number of Advisories, by Fiscal Year and SRO, Fiscal Years 2003-2006

Fiscal year	Number of advisories from NASD[a]	Number of advisories from NYSE	Number of advisories from all other SROs	Total advisories
2003	0	0	5	5
2004	0	1	49	50
2005	0	16	129	145
2006	5	18	167	190
Total	5	35	350	390

Source: GAO.

[a]NASD officials noted that they develop information on unusual market activity as well as they possibly can and typically submit referrals, rather than advisories.

Table 14. Number and Type of Referrals, Fiscal Years 2003-2006

Fiscal year	Number of insider trading referrals	Number of market manipulation referrals	Number of all other types of referrals	Total referrals
2003	283	53	102	438
2004	321	10	9	340
2005	306	24	18	348
2006	386	41	87	514
Total	1,296[a]	128	216	1,640

Source: GAO.

[a]Our analysis shows that from fiscal years 2003 to 2006, almost 80 percent of SRO referrals involved potential insider trading activity, and that almost 60 percent of investigations opened by SEC involved potential insider trading. A SEC branch chief noted that the differences in percentages reflect the difficulty of proving insider trading cases.

Table 15. Number of Referrals, by SRO and Fiscal Year, Fiscal Years 2003-2006

Fiscal year	Number of referrals from NASD	Number of referrals from NYSE	Number of referrals from all other SROs	Total referrals
2003	247	70	121	438
2004	177	39	124	340
2005	130	89	129	348
2006	201	142	171	514
Total	755	340	545	1,640

Source: GAO.

Table 16. Number and Type of Matters Under Inquiry, Fiscal Years 2003-2006

Fiscal year	Number of insider trading MUIs	Number of market manipulation MUIs	Number of all other types of MUIs	Total MUIs
2003	86	40	26	152
2004	147	44	29	220
2005	154	74	37	265
2006	172	89	61	322
Total	559	247	154	960

Source: GAO.

**Table 17. Numbers of Matters Under Inquiry Closed and
Associated Reasons for Closure, Fiscal Years 2003-2006**

Reason for closure	Number of MUIs closed	Percentage of total MUIs closed
Closed into investigation	605	63.0%
Evidence not appropriate for investigation	253	26.4
Closed due to resource limits	38	4.0
Case transferred to another SEC office	29	3.0
Merged with another case	20	2.1
Inappropriate for SRO action	12	1.3
Sent to state or local agency	1	0.1
Sent to SRO for further action	1	0.1
Sent to another federal agency	1	0.1
Total	960	100

Source: GAO.

**Table 18. Number and Type of Investigations Resulting from
SRO Referrals, Fiscal Years 2003-2006**

Fiscal year	Number of insider trading investigations	Number of market manipulation investigations	Number of all other types of investigations	Total investigations
2003	50	17	15	82
2004	89	26	21	136
2005	84	38	26	148
2006	111	60	37	208
Total	334	141	99	574

Source: GAO.

**Table 19. Average and Median Investigation Duration, by Type of
Investigation, Fiscal Years 2003-2006**

Type of investigation	Average duration, by days
All investigations	534
Insider trading	554
Market manipulation	543
All investigations, except insider trading	495

Source: GAO.

**Table 20. Number, Type, and Duration of Investigations,
Fiscal Years 2003-2006**

	Open investigations (as of 4/18/07)		Closed investigations	
Fiscal year	Number	Days of average duration	Number	Days of average duration
2003	36	1,426	46	741
2004	68	1,114	68	565
2005	98	744	50	434
2005	183	372	25	260
Total/Average	385	697	189	534

Source: GAO.

Table 21. Number and Type of Case Actions, Fiscal Years 2003-2006

Fiscal year	Number of insider trading actions	Number of market manipulation actions	Number of all other types of actions	Total actions
2003	2	0	0	2
2004	4	2	2	8
2005	15	5	3	23
2006	13	4	12	29
Total	34	11	17	62

Source: GAO.

Table 22. Number and Type of Case Outcomes, Fiscal Years 2003-2006

Fiscal year	Number of insider trading outcomes	Number of market manipulation outcomes	Number of all other types of outcomes	Total outcomes
2003	3	0	0	3
2004	20	1	2	23
2005	33	4	8	45
2006	40	10	32	82
Total	96	15	42	153

Source: GAO.

REFERENCES

[1] The Securities Exchange Act of 1934 requires SROs to, among other things, be so organized and have the capacity to carry out the purposes of the act and to enforce compliance by its members and persons associated with its members with the rules and regulations of the act and the rules of the SRO. SEC approved the establishment of FINRA in July 2007. FINRA is the result of the consolidation of the former NASD (which regulated the over-the-counter market for exchange-listed and nonexchange-listed securities and provided regulatory services to markets such as the American Stock Exchange and the NASDAQ Stock Market) and the member regulation, enforcement, and arbitration operations of NYSE Regulation, Inc. (NYSE Regulation). However, NYSE Regulation, a subsidiary of NYSE, continues to be responsible for monitoring trading that occurs on NYSE and NYSE Arca, Inc., and conducting investigations of suspicious trades. Because this consolidation occurred after we finished our fieldwork, we refer to the former NASD, and not FINRA, throughout this report.

[2] SEC generally refers to its reviews of SROs, investment companies, and investment advisers as "inspections" and its reviews of registered broker-dealers as "examinations."

[3] On the basis of 2006 data, NYSE and NASD provide market oversight over the two largest exchanges in terms of domestic equity market capitalization and the value of their shares traded.

[4] During a sweep inspection, OCIE probes specific activities of all SROs, or a sample of them, to identify emerging compliance issues.

[5] See GAO, *Securities Regulation: Opportunities Exist to Enhance Investor Confidence and Improve Listing Program Oversight,* GAO-04-75 (Washington, D.C.: Apr. 8, 2004). During our prior review, OCIE officials expressed concern that the routine use of SRO internal audit reports during SRO inspections would have a "chilling effect" on the flow of information between SRO internal audit staff and other SRO employees.

[6] As a result of a 1985 study, SEC determined that SROs had created a viable intermarket surveillance program, and terminated its then tentative Market Oversight and Surveillance System project by determining not to develop the direct surveillance capabilities the system would have allowed. See United States Securities and Exchange Commission, *Final Report to The Senate Committee on Banking, Housing, and Urban Affairs*

and *The House Committee on Energy and Commerce: Regarding the Market Oversight and Surveillance System* (Washington, D.C.: 1985).

[7]   Insider trading is the buying or selling of a security by someone who has access to material, nonpublic information about the security. It is illegal because any trading that is based on this information is unfair to investors who do not have access to the information.

[8]   OCIE also lists NASD district offices as key regulatory programs with routine inspection cycles. OCIE also conducts inspections of other nonexchange SROs, which include registered clearing agencies, transfer agents, and the Municipal Securities Rulemaking Board.

[9]   In addition to FINRA and NYSE, there are nine other SROs that operate or provide regulatory services to an exchange: the American Stock Exchange; the Boston Stock Exchange; the Chicago Board Options Exchange; the Chicago Stock Exchange; the International Securities Exchange; the NASDAQ Stock Market LLC; the National Stock Exchange; NYSE Arca, Inc.; and the Philadelphia Stock Exchange.

[10]   Corrective actions are at times taken prior to the inspection report being issued. In this case, OCIE generally still notes the finding and recommendation in its report.

[11]   Five branch chiefs and 3 assistant directors are located within the Office of Market Oversight.

[12]   Congress passed the Sarbanes-Oxley Act of 2002 in response to corporate failures and fraud that resulted in substantial financial losses to institutional and individual investors. This act substantially increased SEC's appropriations. Pub. L. No. 107-204, 116 Stat. 745 (2002).

[13]   OCIE officials told us that they plan to hire 6 professional staff and 1 branch chief.

[14]   GAO-04-75.

[15]   A requirement for registration as a national securities exchange or national securities association is that the SRO have the capacity to enforce compliance of it members with SRO rules and with the federal securities laws and rules. However, OCIE officials stated that there is no SEC rule that expressly requires SROs to have an internal audit program with prescribed characteristics.

[16]   SEC has recognized that a strong internal audit function contributes to how effectively SROs fulfill their regulatory responsibilities. On at least two occasions, SEC recommended that SROs strengthen this function to improve their oversight. First, an investigation that SEC began in 1994 into the operations and investigations of NASD and the market-making

activities of NASDAQ found that NASD failed over a period to conduct an appropriate inquiry into the anticompetitive actions among NASDAQ market markers. In responding to SEC's resulting recommendations, NASD agreed to ensure the existence of a "substantial" independent review staff reporting directly to NASDAQ's Board of Governors. Second, SEC reported in 1999 that its investigations of the activity of NYSE floor brokers found that NYSE failed to dedicate sufficient resources to allow regulatory staff to perform certain required examinations of floor-broker activity. To address SEC's resulting recommendation, NYSE agreed to maintain its Regulatory Quality Review Department as a "substantial" independent internal review staff with adequate resources to regularly review all aspects of NYSE. (See app. III for additional information on these investigations.)

[17]   SEC enforcement actions and inspections over the past several years have highlighted weaknesses in the effectiveness of certain regulatory programs and raised questions whether, in certain circumstances, SROs have maintained regulatory programs that are sufficiently rigorous to detect, deter, and discipline for member' violations of the federal securities laws and rules and SRO rules. Accordingly, SEC is currently considering the adoption of new rules and the amendment of existing rules designed to provide greater transparency to, among other things, key aspects of the regulatory operations of national securities exchanges and registered securities associations. OCIE officials believe these rules would allow OCIE to better monitor SRO activities between inspections. See *Fair Administration and Governance of Self-Regulatory Organizations, et al.,* 69 Fed. Reg. 71126 (Dec. 8, 2004) (proposed rule).

[18]   SEC's Policy Statement regarding Automated Systems of Self-Regulatory Organization issued in 1989 set for SEC's expectation that SROs establish comprehensive planning and assessment programs to determine the capacity and vulnerability of their IT trading and market information systems. The statement also provides guidance on the components of such a program, which included independent reviews and notification processes for system changes and outages. See *Automated Systems of Self-Regulatory Organizations, Exchange Act Release No. 27445* (Nov. 16, 1989), published in 54 Fed. Reg. 48703 (Nov. 24, 1989). Under the ARP, SEC staff conduct reviews of how SROs are addressing SEC's expectations in these areas. For further information on ARP, see GAO, *Financial Market Preparedness: Significant Progress Has Been Made,*

but Pandemic Planning and Other Challenges Remain, GAO-07-531
(Washington, D.C.: Mar. 29, 2007).

[19] Section 31 of the Securities and Exchange Act requires SEC to collect
transaction fees designed to cover the cost to the government of the
supervision and regulation of the securities markets, including costs
associated with administrative, enforcement, and rulemaking activities. 15
U.S.C. § 78ee.

[20] GAO, *Standards for Internal Control in the Federal Government,*
GAO/AIMD-00.21.3.1 (Washington, D.C.: November 1999).

[21] Between fiscal years 2002 and 2006, OCIE completed an average of 42
inspections of SROs per year.

[22] SEC uses CATS to record key information about MUIs, investigations,
actions, and case outcomes. This information includes basic background
on cases SEC has opened, dates for case milestones, and eventual case
outcomes.

[23] Enforcement officials said that although advisories generally do not
contain enough information to warrant opening an MUI, they found this
sharing of information useful in staying abreast of and potentially
responding to unusual market activity.

[24] Referrals that do not become MUIs are closed, but information on the
referrals still resides in the SRO system. If MUIs approved by OMS
branch chiefs and Enforcement associate directors involve issuers or
individuals in multiple states or in Washington, D.C., MUIs may be
assigned to headquarters Enforcement staff for review and decisions on
whether to fully investigate. Otherwise, branch chiefs assign MUIs to the
appropriate SEC regional office. For example, an MUI that contains
information about suspected insider trading activity among individuals in
a New York firm would be referred to SEC's New York Regional Office.

[25] According to SEC Enforcement officials, SEC's case tracking system
records the beginning of an investigation when Enforcement staff decide
to investigate MUIs and open an investigation. The investigation is
officially closed in the system after administrative or district court
proceedings have concluded and all outcomes, such as fines, other
penalties, and disgorgement, have been collected and distributed. The
investigation average calculated in footnote 23 therefore includes cases
that are filed or instituted as litigated matters, which require additional
time for interim steps, such as discovery depositions and trial. The average
also includes matters where a party is given an extended time in which to
pay disgorgement or penalties, due to his or her financial condition. It also

includes matters where additional noninvestigative time is spent distributing funds to investors through a disgorgement or Fair Fund. The investigation is not formally closed in CATS until all such additional steps are completed.

[26]   The overall referral and investigation processes duration of 726 days, or almost 2 years, consists of a 123-day average for issue identification and SEC referral receipt, 17-day average for SEC to open an MUI, 52-day average for SEC to determine whether to investigate a matter, and 534-day average for SEC to open an investigation and completely conclude a case (see figure 2).

[27]   We calculated the 123-day average duration between SRO issue identification and SEC referral receipt using data from the SRO system on formal referrals. The 123-day average does not include earlier contact by SROs, which may make telephone referrals that may predate formal referrals. In addition, we calculated the 17-day average duration between SEC referral receipt and SEC MUI opening using data on MUIs that SEC opened after receiving referrals from SROs. The 17-day average does not include instances when SEC opened an MUI before receiving an SRO referral.

[28]   We calculated average investigation duration by using 189 of 574 total investigations opened during the period of our review that had open and close dates, and therefore could be used to calculate the average duration. Of the 574 investigations SEC opened during our review period, the remaining 385 (or two thirds) were ongoing or active as of the date SEC provided us with these data (Apr. 18, 2007) and were not used to calculate the 534-day average duration for investigations. We determined that as of this date, these active cases had been open an average of 696 days. Appendix IV provides additional information on these cases.

[29]   Figure 3 is not drawn to scale. Data found in this figure have two sources. The SRO system is the source of data on the number of advisories and referrals, while CATS is the source for the data on MUIs, investigations, actions, and case outcomes.

[30]   Calculating certain durations included in this report required us to manually merge data from the SRO and case tracking systems.

[31]   GAO, Securities and Exchange Commission: Additional Actions Needed to Ensure Planned Improvements Address Limitations in Enforcement Division Operations, GAO-07-830 (Washington, D.C.: Aug. 15, 2007) for more information on CATS management and reporting limitations and

SEC's ongoing efforts to create the Hub to improve Enforcement information system capabilities.

Appendix II

[1]   In July 2007, SEC approved the establishment of the Financial Industry Regulatory Authority (FINRA). FINRA consolidated the former NASD (which provided regulatory services to markets such as the American Stock Exchange and NASDAQ) and the member regulation, enforcement, and arbitration operations of NYSE Regulation. NYSE Regulation, however, continues to be responsible for monitoring trading activity on the NYSE market and conducting investigations of suspicious trades. Because this consolidation occurred after our audit work was complete, we chose to refer to the former NASD, and not FINRA, throughout this report.

[2]   When bluesheeting a broker-dealer, SROs request detailed information about trades performed by the firm and its client, including the stock's name, the date traded, price, transaction size, and a list of the parties involved. The questionnaires SROs use came to be known as blue sheets because they were originally printed on blue paper. Today, due to the high volumes of trades, this information is provided electronically.

[3]   The purpose of the ISG is to provide a framework for the sharing of information and the coordination of regulatory efforts among exchanges trading securities and related products to address potential intermarket manipulations and trading abuses.

[4]   Referrals from SROs grew from 438 to 514, or an increase of 17 percent, between fiscal years 2003 and 2006. The numbers of SEC investigations and enforcement actions also showed a corresponding increase. We found that almost 91 percent of all advisories and almost 80 percent of SRO referrals sent to SEC during this period involved suspected insider trading activity, which Enforcement and SRO staff attributed to increased merger and acquisition activity.

[5]   A short sale is the sale of a borrowed security, commodity, or currency with the expectation that the asset will fall in value. For example, an investor who borrows shares of stock from a broker and sells them on the open market is said to have a short position in the stock. The investor must eventually return the borrowed stock by buying it back from the open market. If the stock falls in price, the investor buys it for less than he or she sold it, thus making a profit.

[6]   *Order Approving Options Regulatory Surveillance Authority Plan,* Exchange Act Release No. 34-53940 (June 5, 2006), published in 71 Fed. Reg. 34399 (2006) (Order).

Chapter 2

ADDITIONAL ACTIONS NEEDED TO ENSURE PLANNED IMPROVEMENTS ADDRESS LIMITATIONS IN ENFORCEMENT DIVISION OPERATIONS*

ABBREVIATIONS

AIG	American International Group, Inc.
CATS	Case Activity Tracking System
ERISA	Employee Retirement Income Security Act of 1974
FINRA	Financial Industry Regulatory Authority
MUI	matter under inquiry
OCIE	Office of Compliance Inspections and Examinations
OIT	Office of Information Technology
OMB	Office of Management and Budget
OMS	Office of Market Surveillance
SEC	Securities and Exchange Commission
SRO	self-regulatory organization

---

* Excerpted from CRS Report GAO-07-830, dated August 2007.

August 15, 2007
The Honorable
Charles E. Grassley
Ranking Member
Committee on Finance United States Senate

Dear Senator Grassley:

The Securities and Exchange Commission's (SEC) ability to conduct investigations and bring enforcement actions for violations of securities laws is critical to its mission to protect investors and maintain fair and orderly markets. SEC's Division of Enforcement (Enforcement) is charged with investigating securities law violations; recommending civil enforcement actions when appropriate, either in a federal court or before an administrative law judge; and negotiating settlements on behalf of the Commission. The types of sanctions that Enforcement can seek on behalf of the Commission include monetary penalties or fines and disgorgements of the profits that individuals or companies may derive by having committed securities violations.[1] While SEC has only civil authority, it also works with various law enforcement agencies, including the United States Department of Justice (Justice), to bring criminal cases when appropriate. In addition, Enforcement is responsible for overseeing the Fair Fund program, which seeks to compensate investors who suffer losses resulting from fraud or other securities violations by individuals and companies.[2] Under the Fair Fund program, SEC can combine the proceeds of monetary penalties and disgorgements into a single fund and then distribute the proceeds to harmed investors.

In recent years, Enforcement has initiated high-profile actions that resulted in record civil fines against companies and senior officers and in some cases contributed to criminal convictions.[3] However, the capacity of SEC in general and Enforcement in particular to appropriately plan and effectively manage their activities and fulfill their critical law enforcement and investor protection responsibilities on an ongoing basis has been criticized in the past. Although SEC received a substantial increase in its appropriations as a result of the Sarbanes-Oxley Act of 2002, questions have been raised in Congress and elsewhere on the extent to which the agency is using these resources to better fulfill its mission.[4] Moreover, we have reported that aspects of Enforcement's information systems and management procedures could limit the efficiency and effectiveness of its operations.[5] For example, we found in 2004 that Enforcement faced challenges in developing the advanced information technology necessary to facilitate the investigative process.[6] In addition, we reported in 2005 that the distribution of funds to harmed investors under the Fair Fund program was limited and that Enforcement had not developed adequate systems and data to fulfill its oversight responsibilities.[7]

Because of your interest in ensuring that SEC effectively manages its resources and helps ensure compliance with securities laws and regulations, you requested that we review key Enforcement management processes and systems and follow up on our previous work where appropriate. Accordingly, this chapter evaluates Enforcement's (1) internal processes and information systems for planning, tracking, and closing investigations and planned changes to these processes and systems; (2) implementation of SEC's Fair Fund program responsibilities; and (3) efforts to coordinate investigative activities with other SEC divisions and federal and state law enforcement agencies.

To address all three objectives, we obtained and reviewed relevant SEC and Enforcement documentation and data. Specifically, we reviewed documentation and data relating to Enforcement's planning processes; its automated system for tracking investigations and enforcement actions—the Case Activity Tracking System (CATS)—and a planned successor system; the Fair Fund program; and internal and external coordination.[8] We also reviewed our relevant prior reports and federal standards for internal controls. Further, we interviewed the SEC Chairman and two commissioners, senior agency and Enforcement officials in Washington, and officials in three SEC regional offices (Boston, New York, and Philadelphia) that are responsible for a significant share of Enforcement's investigative activity. We also contacted Enforcement officials in other SEC offices as appropriate. Additionally, we interviewed consultants that assist Enforcement in developing plans to distribute funds to harmed investors under the Fair Fund program.

We conducted our work in Washington, D.C., Boston, Massachusetts, New York, New York, and Philadelphia, Pennsylvania, between November 2006 and July 2007 in accordance with generally accepted auditing standards. Appendix I explains our scope and methodology in greater detail.

RESULTS IN BRIEF

Enforcement's processes and systems for planning, tracking, and closing investigations have had some significant limitations that have hampered the division's capacity to effectively manage its operations and allocate limited resources. While Enforcement and SEC officials are aware of these deficiencies and have recently begun addressing them, additional actions are necessary to help ensure that the planned improvements fully address limitations in the division's operations. The following points summarize key issues:

- In March 2007, Enforcement said it would centrally review and approve all new investigations of potential securities law violations by individuals or companies. Under Enforcement's previous, largely decentralized approach (senior Enforcement attorneys in the agency's home and 11 regional offices could approve new investigations), the division was not always able to ensure the efficient allocation of resources or maintain quality control in the investigative process. While the new centralized approach was designed to help address these issues, Enforcement has not yet established written procedures and criteria for reviewing and approving new investigations. Without such procedures and criteria, Enforcement may face challenges in consistently communicating the new approach to existing and new staff. The lack of written procedures and criteria could also limit the Commission's ability to evaluate the implementation of the new approach and help ensure that the division is managing its operations and resources efficiently.

- Recognizing that the division's current information system for tracking investigations and enforcement actions—CATS—is severely limited as a management tool, Enforcement plans to start using a new system (the Hub) by late 2007. The deficiencies of CATS include its inability to produce detailed reports on investigations of certain types (for example, those for hedge funds) or the status of such investigations.[9] While the Hub is designed to address many of CATS's deficiencies—it will, for example, be able to produce detailed management reports on ongoing investigations— the way that the system is being implemented may not address all existing limitations. More specifically, Enforcement has not established written controls to help ensure that staff enter investigative data in the Hub in a timely and consistent manner. Without such controls, management reports generated by the Hub may have limited usefulness, and the system's capacity to assist Enforcement in better managing ongoing investigations will not be fully realized.

- In May 2007, Enforcement implemented procedures to help ensure the prompt closure of investigations that are no longer being pursued and thereby better ensure the fair treatment of individuals and companies under review, but these procedures do not fully address the entire backlog of these investigations. One regional Enforcement official said that as of March 2007, nearly 300 (about 35 percent) of the office's 840 open investigations were 2 or more years old, were no longer being pursued, and had no pending enforcement actions.[10] Enforcement officials said that the failure to close such investigations promptly could have negative consequences for individuals and companies no longer suspected of having committed securities violations. They attributed the failure to close many investigations to several factors, such as time-consuming administrative requirements for attorneys to prepare detailed investigation closing memorandums that then must be routed to senior division officials for review and approval. To address these issues, Enforcement plans to inform individuals and companies more promptly that they are no longer under review and expedite the review and closure of the existing backlog of investigations for which administrative tasks have been completed (as of March 2007, there were 464 such investigations). However, Enforcement's plans do not include clearing the potentially large backlog of investigations for which such administrative tasks have not been completed, which could be negatively impacting individuals and companies no longer actively under review.

Enforcement's management of the Fair Fund program may have contributed to delays in distributing funds to harmed investors, and the division lacks data necessary for effective program oversight. For the 115 Fair Funds currently tracked by Enforcement (which were created by federal courts or through SEC administrative proceedings), only about $1.8 billion (about 21 percent) of the $8.4 billion ordered since the program's inception in 2002 had been distributed to harmed investors as of June 2007, according to SEC data.[11] Enforcement officials and consultants who administer Fair Fund plans have attributed the limited payout rate to factors such as difficulties in identifying harmed investors, the complexity of individual cases, and the need to resolve related tax issues. However, Enforcement's largely decentralized approach to managing the Fair Funds program may have also contributed to distribution delays. While senior Enforcement officials in Washington have a coordination and oversight role, staff attorneys in either the home or the regional offices that brought the related enforcement action are primarily responsible for overseeing consultants who design and execute Fair Fund distribution plans. However, this delegated management structure appears to have impeded the development of uniform Fair Fund procedures that otherwise could have facilitated the distribution of funds to harmed investors. In addition, Enforcement officials said that the management structure diverts investigative attorneys from their primary law enforcement mission. In response to these concerns, in March 2007 SEC's Chairman announced a plan to centralize the administration of the Fair Fund program within a new office. However, it is too soon to assess how this new office will affect the program because SEC has not yet staffed the office or developed written guidance to define its role, responsibilities, and procedures. Moreover, Enforcement does not yet systematically collect or analyze key Fair Fund data, such as the administrative expenses that consultants are incurring to design and execute Fair Fund plans, as we recommended in 2005.[12] While Enforcement officials agree that reviewing such data would enhance their capacity to assess the reasonableness of Fair Fund administrative costs, an information system designed to collect and report such expense data for ongoing plans is not expected to be completed until 2008. In the meantime, Enforcement has not ensured that reports intended to provide expense data for completed Fair Fund plans contain consistent information or are analyzed.[13] Without such information, Enforcement's Fair Fund oversight capacity is limited.

Enforcement coordinates investigations and other activities with other SEC divisions and outside law enforcement authorities and is implementing our previous recommendation that it document referrals to criminal investigative authorities. According to Enforcement officials, they regard coordinating the division's investigative activities with SEC's Office of Compliance Inspections and Examinations (OCIE) as particularly important because OCIE staff regularly examine regulated entities and have a broad understanding of the extent of their compliance with laws and regulations. However, Enforcement officials historically have been concerned that OCIE referrals lacked sufficient information. As a result, Enforcement and OCIE have recently instituted a new committee process to formally review such referrals and track their outcome. Further, Enforcement officials said that the division has established working relationships with U.S. attorney offices and state securities regulators to leverage investigative resources. Enforcement also held coordination conferences attended by federal and state agencies. However, Enforcement is in the process of implementing our 2005 recommendation to document informal referrals of potential criminal matters, which it intends to do through the planned investigation and enforcement action tracking system—the Hub.[14] Until the system is in place, Enforcement cannot readily determine and verify whether staff make appropriate and prompt referrals, and the division lacks an institutional record of the types of matters that have been referred over the years. Without such information, Enforcement's ability to manage and oversee the referral process is limited.

This chapter makes several recommendations designed to strengthen Enforcement's management of the investigation process and the Fair Fund program. In brief, the report recommends that the SEC Chairman direct Enforcement and other agency offices, as appropriate, to (1) establish written policies and assessment criteria for reviewing and approving new investigations, (2) establish controls to better ensure the reliability of investigative data entered into the Hub information system, (3) consider developing expedited procedures for closing investigations, and (4) establish a comprehensive plan to staff and identify the roles and responsibilities of the new Fair Fund program office and collect and analyze reports on completed Fair Fund plans.

We provided a draft of this chapter to SEC, and the agency provided written comments that are reprinted in appendix III. In its written comments, SEC agreed with our conclusions and stated that it would implement all of our recommendations. Moreover, SEC officials noted that the agency has since established that the new Fair Fund office—referred to as the Office of Distributions, Collections and Financial Management—will be located within the Division of Enforcement. SEC said that a senior officer and two assistant directors will lead the operations of the office and the agency is developing the office's responsibilities. SEC also provided technical comments, which we have incorporated as appropriate.

BACKGROUND

SEC is an independent agency created in 1934 to protect investors; maintain fair, honest, and efficient securities markets; and facilitate capital formation. The agency has a five-member Commission that the President appoints, with the advice and consent of the Senate, and that a Chairman designated by the President leads. The Commission oversees SEC's operations and provides final approval of SEC's interpretation of federal securities laws, proposals for new or amended rules to govern securities markets, and enforcement activities. Table 1 identifies several key SEC units and summarizes their roles and responsibilities.

SEC's current 2004-2009 strategic plan established four goals: (1) enforce compliance with the federal securities laws, (2) promote healthy capital markets through an effective and flexible regulatory environment, (3) foster informed investment decision-making, and (4) maximize the use of SEC resources. Enforcement and OCIE share joint responsibility for implementing the agency's first strategic goal. The Commission and the Office of the Executive Director, which develops and implements all the agency's management policies, are updating the agency's strategic plan, which is to be issued in the summer of 2007.

Enforcement personnel are located in SEC's home office in Washington, D.C., as well as the agency's 11 regional offices.[15] Enforcement staff located in the home office include the director and one of two deputy directors, five investigative groups or Offices of Associate Directors, as well as internal support groups, including its Offices of Chief Counsel and Chief Accountant (see figure 1).[16] An associate director heads each Office of Associate Director and has one or more assistant directors. Branch chiefs report to assistant directors and supervise the work of investigative staff attorneys assigned to individual investigations, with review and support provided by division management. SEC regional office staff are typically divided between Enforcement and OCIE personnel. Enforcement units in the regional offices have Office of Associate Director structures similar to those in the home office and report to the Director of Enforcement in Washington, D.C.

Table 1. Roles and Responsibilities of SEC Divisions and Offices

Division or office	Roles and responsibilities
Division of Enforcement	Conducts investigations of registered entities (such as broker-dealers and investment advisers) or unregistered entities (such as unregistered and fraudulent securities offerings over the Internet), recommends Commission action (either in a federal court or before an administrative law judge), negotiates settlements on behalf of the Commission, and works with criminal law enforcement agencies when warranted.
Division of Corporation Finance	Reviews corporate disclosures, assists companies in interpreting the Commission's rules, and recommends new rules for adoption.
Division of Market Regulation	Establishes and maintains standards for fair, orderly, and efficient markets by regulating the major securities market participants, including broker-dealers, self-regulatory organizations (SRO), transfer agents (parties that maintain records of stock and bond owners), and securities information processors.[a]
Division of Investment Management	Regulates the investment management industry and administers the securities laws affecting investment companies and advisors.
Office of the Chief Accountant	Establishes and enforces accounting and auditing policy to enhance financial reporting and improve the professional performance of public company auditors.

Division or office	Roles and responsibilities
Office of Compliance Inspections and Examinations	Administers a nationwide examination and inspection program for registered SROs, broker-dealers, transfer agents, clearing agencies, and investment companies and advisors to quickly and informally correct compliance problems.
Office of Economic Analysis	Serves as the chief advisor to the Commission and its staff on all economic issues associated with the SEC's regulatory activities, analyzes the likely consequences of proposed regulations, and engages in research to support longer term SEC policy initiatives and plans.
Office of General Counsel	Represents SEC in various proceedings, prepares legislative materials, and provides independent advice and assistance to the Commission, divisions, and offices.
Office of Investor Education and Assistance	Provides information to investors, seeks informal resolutions of complaints, and collects data on investor contacts to track trends in the security industry and provide intelligence to other SEC divisions and offices.

Source: SEC.

[a] SROs include national securities exchanges (stock exchanges), the Financial Industry Regulatory Authority (FINRA), and clearing agencies, which facilitate trade settlements. FINRA was created in July 2007 through the consolidation of NASD (formerly an SRO) and the member regulation, enforcement, and arbitration functions of the New York Stock Exchange; it is now the largest nongovernmental regulator for all securities firms doing business in the United States.

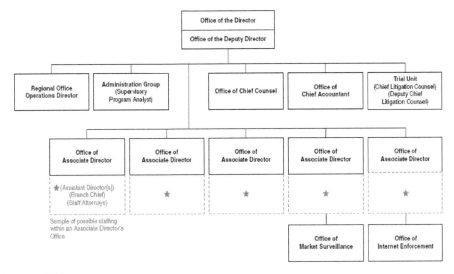

Source: SEC.

Figure 1. Organizational Structure of SEC's Enforcement Division Home Office.

The Sarbanes-Oxley Act of 2002 substantially increased SEC's appropriations, and Enforcement subsequently increased its staffing levels. In 2002, Enforcement had 1,012 staff and, at the end of fiscal year 2006, 1,273 staff.[17] As shown in figure 2, the number of investigative attorneys in Enforcement increased substantially, from 596 in 2002 to 740 in 2005.[18] However, the number of staff in Enforcement, in particular its investigative attorneys, decreased from 2005 to 2006 because of a May 2005 hiring freeze (instituted across the agency in response to diminished budgetary resources) and subsequent attrition. Since October 2006, however, SEC has permitted Enforcement and other SEC divisions and offices to replace staff that leave the agency. However, the agency does not contemplate returning to early 2005 staffing levels. Appendix II provides additional information on Enforcement's staffing resources and workload indicators.

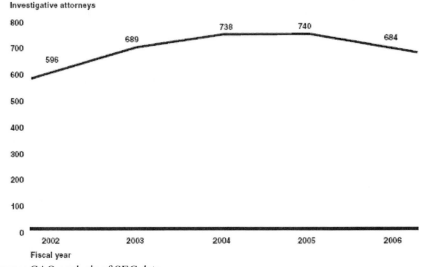

Source: GAO analysis of SEC data.

Figure 2. Number of Investigative Attorneys at Fiscal Year End, 2002-2006.

Figure 3 provides a general overview Enforcement's investigative process. At the initial stage of the investigative process, attorneys evaluate information that may indicate the existence of past or imminent securities laws violations. The information can come from sources such as tips or complaints from the public as well as referrals from other SEC divisions or government agencies. If Enforcement staff decide to pursue the matter, they will open either a Matter

Under Inquiry (MUI) or an investigation. Staff open a MUI when more information is required to determine the merits of an investigation; otherwise, staff may open an investigation immediately.[19] Investigations can be conducted informally—without Commission approval—or formally, in which case the Commission must first approve a formal order if staff find it necessary to issue subpoenas for testimony or documentation. Based on the analysis of collected evidence, Enforcement will decide whether or not to recommend that the Commission pursue enforcement actions, which can be administrative or federal civil court actions (both of which must be authorized by the Commission). Enforcement has established a variety of controls over the enforcement action process, including reviews by senior division officials in Washington, D.C., and, ultimately, review and approval by the Commission.[20] Enforcement has an information technology system—CATS—that tracks the progress of its MUIs, investigations, and enforcement actions.

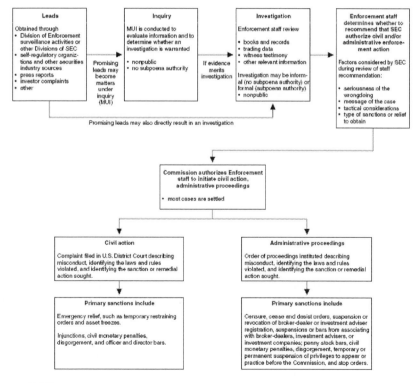

Source: GAO.

Figure 3. Flowchart of SEC's Investigation and Enforcement Process.

Enforcement also is responsible for implementing and overseeing the Sarbanes-Oxley Act's Fair Fund provision, which allows SEC to combine civil monetary penalties and disgorgement amounts collected in enforcement cases to establish funds for investors harmed by securities laws violations.[21] Fair Funds may be created through either SEC administrative proceedings or litigation in U.S. District Court, and either SEC or the courts may administer the funds. However, SEC is responsible for general monitoring of all Fair Funds created. Typically, for SEC-ordered Fair Funds, the agency hires consultants to create Fair Fund distribution plans (independent distribution consultants) and oversee payments to harmed investors (fund administrators). However, in some cases, SEC staff will take care of all of the distribution responsibilities internally. The development of a Fair Fund plan can include estimating losses suffered by harmed investors. For court-ordered funds, SEC recommends a receiver or distributions agent, who creates a distribution plan that is presented for court approval.

ENFORCEMENT HAS TAKEN STEPS TO BETTER MANAGE THE INVESTIGATIVE PROCESS, BUT THESE STEPS MAY NOT FULLY ADDRESS EXISTING LIMITATIONS

Enforcement's approaches for planning, tracking, and closing investigations have had some significant limitations that have hampered its ability to effectively manage its operations, allocate limited staff resources, and ensure the fair treatment of individuals and companies under investigation. While SEC and Enforcement officials are aware of these limitations and have begun addressing them, some of their actions may not fully correct identified weaknesses. Specifically, Enforcement has not (1) established written procedures and criteria for its newly centralized review and approval process for new investigations, which could limit its ability to ensure its consistent implementation and reduce the Commission's ability to oversee the division's operations; (2) established controls to help ensure the reliability of the investigative data that division attorneys will be required to enter into a new information system, which could limit the usefulness of management reports generated by the system; and (3) established plans and procedures to ensure that all investigations that are no longer being actively pursued are closed promptly to reduce the negative impact on individuals and companies no longer under review.

Enforcement Recently Centralized the Review and Approval of New Investigations, but the Lack of Written Procedures and Criteria Could Limit the Effectiveness of Its Approach

To establish overall investigative priorities, Enforcement officials said that they regularly communicate with senior SEC officials and their counterparts in other agency units. For example, Enforcement officials said that they hold weekly meetings with the SEC Chairman and other commissioners as appropriate. During the Chairman's tenure, he has identified the pursuit of securities fraud against senior citizens as a key investigative priority for Enforcement and other agency offices, including OCIE.[22] In addition to specific priorities, Enforcement officials said that they seek to maintain a constant investigative presence across all areas of potential securities violations (for example, insider trading abuses) and that this "cover the waterfront" approach is designed to prosecute and possibly deter securities law offenders. While an Enforcement official said that the division has not established minimum quotas for different types of investigations and enforcement actions, it will intervene if any one type threatens to overwhelm the division's operations. Based on internal analysis of enforcement action data, Enforcement officials determined that if the division's pursuit of any type of securities enforcement action exceeded 40 percent of total enforcement actions, an unacceptable commitment of division resources would result.[23]

While Enforcement has established planning processes for determining overall priorities, the division has used a largely decentralized approach for reviewing and approving individual new investigations, which may have limited the division's operational effectiveness, according to senior SEC and Enforcement officials. Under this traditional approach, associate directors in either SEC's home or 11 regional offices approved the opening of MUIs after staff came across a potential violation of federal securities law.[24] While Enforcement's senior leadership in the home office reviewed proposals for formal investigations and received weekly reports on MUIs and new investigations that had been approved in each office (and reviewed summaries of all investigations on a quarterly basis), it did not have formal approval responsibility for such new MUIs and investigations. According to Enforcement officials, staff in each office generally decided to open MUIs and investigations based on considerations such as the likelihood that they would be able to find and prove a violation of federal securities laws, the potential amount of investor loss, the gravity of the misconduct, and the potential message the case would deliver to the industry and public. Typically, the staff attorney who opened the MUI was responsible for conducting the investigation. According to Enforcement officials, this decentralized approach was generally viewed as fostering creativity in the investigative process and providing staff with incentives to actively seek potential investigations.

However, without a centralized control mechanism for reviewing and approving all new MUIs and investigations, Enforcement's capacity to ensure the efficient use of available resources, which is one of SEC's four strategic goals, was limited. For example, SEC's Chairman, officials from his office and the Office of the Executive Director, and Enforcement officials said that the division has not always been able to prioritize or ensure an efficient allocation of limited investigative staff resources. Officials said that in some cases staff attorneys worked on investigations that were outside of their geographic area (for example, San Francisco staff conducting an investigation in the Atlanta region). Consequently, the officials said that the division incurred travel and other related costs that could have been minimized if a centralized process had been in place to approve all new investigations. Further, one official from the Chairman's office said that without a formal quality check by senior Enforcement officials, in some cases MUIs and investigations had been opened and allowed to linger for years with little likelihood of resulting in enforcement actions.

In March 2007, Enforcement began using a new, more centralized approach to review and approve investigations. Under the new approach, two deputy directors, who report directly to the Director of Enforcement, are to review and approve all newly opened MUIs and investigations to ensure the appropriateness of resource allocation considerations and whether the MUI should be pursued. One deputy director is to review MUIs opened in the division's home office and another deputy director, based in New York, is to review MUIs opened in regional offices. In addition to the MUI review, after an investigation is open for 6 months, staff will be required to prepare a memorandum with information on evidence gathered to date, whether an enforcement action is likely, resources, and estimated time frames for review by their deputy director. According to Enforcement officials, the goal of this new approach is to provide early assessments of whether an investigation ought to be pursued further and resources reallocated. The deputy directors are also expected to use this review to determine if the investigation is being conducted in a timely manner, if it should be reprioritized based on Enforcement's current caseload, or if it should be closed.

While these are positive developments, Enforcement has not yet established comprehensive written policies specifying how the new approach will be carried out or the criteria that will be used to assess new MUIs and ongoing investigations. According to our and Office of Management and Budget (OMB) standards, documentation is one type of control activity that will help ensure that management's directives, such as these new procedures, are carried out.[25] In spring 2007, Enforcement developed and distributed divisionwide a one-page planning document that, among other items, identified the new centralized approach for reviewing and approving MUIs and investigations. However, without the establishment of agreed-upon and written procedures for carrying out the new approach and relevant assessment criteria, the division may face challenges in consistently communicating and explaining the new approach to all current and new staff. Moreover, the Commission's ability to oversee how effectively Enforcement is implementing the new approach and generally managing its operations may be limited. For example, the lack of a transparent and documented standard could limit the Commission's capacity to identify inconsistencies in the implementation of the new approach, determine whether any such inconsistencies have affected Enforcement's operations, and take corrective action as warranted.

Planned Investigative Information System Has Significant Potential Benefits, but Enforcement Has Not Taken Sufficient Steps to Help Ensure Data Reliability

Enforcement officials have consistently stated that the division's current information system for tracking investigations and enforcement actions—CATS—is severely limited and virtually unusable as a management tool. In particular, the officials have said that access to CATS is limited and the system does not allow division management to generate summary reports, which could be used to help manage operations on an ongoing basis. Currently, the only summary reports CATS readily produces for management review are lists of all open MUIs, investigations, and enforcement actions by general violation types, such as violations involving broker-dealers or investment advisers. CATS does not allow its users to create timely reports on more specific topics, such as ongoing investigations involving hedge funds, which do not exist as classification fields in the system. As a result of the system's limitations, several senior Enforcement management officials said that they maintain their own manual lists of certain types of investigations (such as those for hedge funds) to assist in managing division activities.

Further, to obtain customized reports and statistics on Enforcement operations, division officials said that they must submit requests to SEC's Office of Information Technology (OIT) and then wait for OIT staff to create and run a computer program to respond to the request. Enforcement officials said that OIT staff generally are responsive and work very hard to address these requests; however, given their heavy workload, one Enforcement official said that it generally takes 1-2 days to receive the information, and more complex requests can take as long as a week. Further, Enforcement officials said that obtaining technical support for CATS can be difficult because the system is proprietary and the company that created it is no longer in business. According to Enforcement officials, CATS's deficiencies result from the fact that the system was hastily designed in preparation for expected year 2000 technical challenges.

Having recognized CATS's limitations, SEC and Enforcement officials are developing a new investigation information management system, called the Hub, which is scheduled to be in use divisionwide by the end of fiscal year 2007. According to Enforcement officials, division officials and staff in SEC's Boston office developed a prototype of the Hub in 2004 because of their dissatisfaction with CATS. Subsequently, Enforcement, in coordination with OIT, developed an enhanced version of the Hub, which was then tested among home and regional office staff in late 2006 and early 2007. Enforcement officials said that the Hub is an interim system that will continue to interface with the CATS database until the second phase of the Hub fully replaces CATS, which is expected to occur in fiscal year 2009.

Although the Hub is an interim system, Enforcement officials said that it will significantly enhance the division's capacity to manage the investigative process. In particular, the officials said that the Hub will facilitate the creation of a variety of management reports on the division's investigative activities, including detailed reports on ongoing investigations by certain types (for example, reports on the number of hedge fund investigations). The Hub will also provide more detailed information on the status of investigations so management can better track their progress and timeliness. Further, the officials said that the Hub is designed to be (1) generally accessible to all division staff, although highly sensitive investigative information will be restricted on a need-to-know basis; (2) user-friendly, primarily employing drop-down menus for data entry; (3) searchable so that staff can identify relevant information associated with an investigative matter; and (4) flexible, because new data fields can be added. We reviewed prototype screens for the Hub and found that they were consistent with the descriptions of Enforcement officials, and staff we contacted generally made favorable comments about the system.

However, due to significant planned changes to Enforcement's traditional approach for recording investigative data, there is a risk that data may not be entered into the Hub on a timely and consistent basis, as required by federal internal control standards.[26] Enforcement has traditionally required support personnel or case management specialists (rather than attorneys) to enter investigative data into CATS because of the limited access to the system and its lack of user friendliness. However, once the Hub is implemented in late 2007, Enforcement officials said that they plan to require division attorneys to enter relevant data into the system for all investigations opened after that date. Further, Enforcement officials said that attorneys will be responsible for entering relevant data into the Hub for ongoing investigations that are being actively pursued but were initiated prior to the system's implementation.[27] Enforcement officials regard the entry of such data as critical; otherwise, management reports generated by the Hub would only include information on investigations begun after the system's scheduled implementation in late 2007. One Enforcement official said that the decision to require attorneys to enter data into the Hub was based on the view that such attorneys have firsthand knowledge of ongoing investigations and thus would be able to streamline the process. However, another Enforcement official said that requiring attorneys to maintain timely, accurate, and consistent investigative data in the Hub would require a cultural change on the attorneys' part because they have become accustomed to relying on case management specialists to perform this task. Another Enforcement official questioned whether division attorneys would enter investigative data into the Hub on a timely and consistent basis because they may view doing so as another administrative requirement diverting them from their primary investigative responsibilities.

While Enforcement's plans to require attorneys rather than case management specialists to enter data into the Hub may be appropriate, the division plans only a limited number of actions to ensure that data entered into the system are timely, consistent, and reliable. For example, Enforcement plans to train attorneys on the Hub as it is implemented and is developing a system user manual. However, Enforcement is not developing written guidance identifying data entry into the Hub as a priority for division attorneys and specifying how and when such data entry is to be done. Moreover, Enforcement has not yet established a written process that would allow division officials to independently review and determine the extent to which data entry for the Hub is performed on a timely, consistent, and reliable basis in accordance with federal internal control standards.[28] Without doing so, the usefulness of management reports generated by the Hub may be limited, and the system's potential to significantly enhance Enforcement's capacity to better manage the investigative process may not be fully realized.

**Enforcement Has Planned Improvements to Its Investigation
Closure Processes, but Plans May Not Fully
Address Backlog of Cases**

Enforcement may leave open for years many investigations that are not being actively pursued with potentially negative consequences for individuals and companies no longer under review. According to CATS data, about two-thirds of Enforcement's nearly 3,700 open investigations as of the end of 2006 were started 2 or more years before, one-third of investigations at least 5 years before, and 13 percent at least 10 years before. According to an Enforcement official, technical limitations in CATS make it difficult to readily determine how many of these investigations resulted in enforcement actions and how many did not.[29] Nevertheless, other data suggest that the number of aged investigations that did not result in an enforcement action may be substantial. For example, Enforcement officials at one SEC regional office said that as of March 2007, nearly 300 of 841 open investigations (about 35 percent) were more than 2 years old, had not resulted in an enforcement action, and were no longer being actively pursued.

Enforcement officials cited several reasons for division attorneys not always closing investigations promptly. In particular, the officials said that Enforcement attorneys may view pursing potential securities violations as the division's highest priority and lack sufficient time, administrative support, and incentives to comply with established administrative procedures for closing investigations. For example, Enforcement requires attorneys to complete closing memoranda for each investigation that is to be closed. These memoranda must identify why the investigation was opened, describe the work performed, and detail the reasons for recommending that the investigation be closed without an action. Staff must also prepare draft termination letters, which inform individuals or companies that they are no longer under review. A closing memorandum is also required for investigations with associated enforcement actions. In these cases, the staff attorney must account for all ordered relief before the investigation is closed. One regional Enforcement official estimated that it could take as long as a month for a staff attorney to complete this process and submit the closing package to the home office, although senior division officials noted that attorneys typically would not spend all their time doing so. Once closing packages are received by the home office, Enforcement's Office of Chief Counsel must then approve the closing of the investigation, at which point final termination letters are sent to affected individuals and companies.

Enforcement officials in SEC's home office said that a lack of resources in their office also contributed to delays in closing investigations. They noted that only one person in the division was assigned to processing closing packages for investigations. Consequently, the officials said there was a backlog of investigations for which the closing package had been completed but not reviewed. As of March 1, 2007, the backlog consisted of 464 investigations, according to an Enforcement official.

However, Enforcement officials told us that in May 2007 they began eliminating the backlog of investigations with completed—but unreviewed—closing packages and had almost eliminated the backlog by mid-June 2007. The division recently added one staff person to work on administering closing procedures in the home office, and Enforcement officials have set a goal of processing new closing documentation within 2 weeks of receipt.

Also in May 2007, Enforcement implemented revised procedures for sending termination letters for investigations that will not result in an enforcement action. Under the procedures, Enforcement will send the letters to individuals and companies at the start of the closing process rather than at the end. This particular effort will be emphasized on Enforcement's intranet—EnforceNet. Enforcement officials said they changed this procedure out of concerns about fairness to those under investigation and to reduce any negative impact an open investigation may have on them. For example, a company may bar an individual from performing certain duties until a pending SEC investigation is resolved. Staff are generally encouraged to close investigations if they know they will not be bringing any enforcement actions, even if all of their investigative steps have not yet been completed.

While the above steps are a positive development, they do not address the potentially large backlog of investigations that are not likely to result in enforcement actions and for which closing packages have not been completed. As a result, the subjects of many aged and inactive investigations may continue to suffer adverse consequences until closing actions are completed. We recognize that reviewing and resolving this potentially large backlog of investigations and enforcement actions likely would impose resource challenges for Enforcement. Nevertheless, the failure to address this issue—potentially through expedited administrative closing procedures for particularly aged investigations—would limit Enforcement's capacity to manage its operations and ensure the fair treatment of individuals and companies under its review.

ENFORCEMENT'S MANAGEMENT OF FAIR FUNDS MAY HAVE CONTRIBUTED TO DISTRIBUTION DELAYS, AND THE DIVISION LACKS DATA NECESSARY FOR EFFECTIVE PROGRAM OVERSIGHT

According to available SEC data, the distribution of funds to harmed investors under the Fair Fund program remains limited after 5 years of operation. Enforcement officials, as well as consultants involved in Fair Fund plans, have cited a variety of reasons for the slow distribution, including challenges in identifying harmed investors, the complexity of certain Fair Funds, and the need to resolve tax and other issues. However, the largely decentralized approach that Enforcement and SEC have used in managing the Fair Fund program may also have contributed to distribution delays. SEC has announced plans to create a central Fair Funds office, but it is too early to assess this proposal, as final determinations about its staffing, roles and responsibilities, and procedures have not yet been determined. Further, Enforcement does not yet collect key data necessary to effectively monitor the Fair Fund program (such as data on fund administrative expenses for ongoing plans) because an information system designed to capture such data is not expected to be implemented until 2008. In the meantime, Enforcement has not ensured that reports intended to provide expense data for completed Fair Fund plans contain consistent information or are analyzed. Until Enforcement clearly defines the Fair Fund office's oversight roles and responsibilities and officials establish procedures to consistently collect and analyze additional data, the division will not be in an optimal position to help ensure the effective management of the Fair Fund program.

Fair Fund Distributions Have Been Limited

As of June 2007, Enforcement officials said that they were tracking 115 Fair Funds created since the program's inception in 2002—up from the 75 identified in our 2005 report—largely because funds were created as part of a series of enforcement actions against mutual fund companies.[30] The Fair Fund plans vary considerably in size and complexity, ranging from plans for small broker-dealers with relatively few customers to large corporate cases, according to SEC data. The smallest Fair Fund plan established (measured by the amount of funds ordered returned to investors) was $29,300 for alleged fraud at a hedge fund; the largest was $800 million for alleged securities fraud at American International

Group, Inc. (AIG). Table 2 shows the 10 largest Fair Funds ordered through June 2007; 7 are court-created plans, and 3 have been established through SEC administrative proceedings. SEC monitors all Fair Fund plans regardless of their source.

According to SEC data, from 2002 to 2007, federal courts and SEC administrative proceedings ordered individuals and entities subject to SEC enforcement actions to pay a total of $8.4 billion into Fair Fund plans, an increase of about 75 percent from the $4.8 billion total Fair Funds we identified in our 2005 report. As of June 2007, $1.8 billion of the $8.4 billion (about 21 percent) had been distributed to harmed investors, according to SEC data. As shown in table 3, the amount distributed from court-overseen plans exceeded that distributed from SEC-overseen plans. According to Enforcement officials, the funds were distributed more slowly from SEC-overseen plans largely because much of the money ordered through SEC proceedings involves mutual fund market timing matters, which, as discussed later, are among the most complex Fair Fund plans.

Table 2. The 10 Largest Fair Funds Ordered, as of June 2007

Fair Fund	Alleged type of activity	Source	Judgment date	Total ordered
AIG	Improper accounting and workers' compensation practices	Court	2/17/2006	$800,000,000
Worldcom	Overstating income	Court	7/7/2003	$750,000,000
Wall Street research analysts	Research and investment banking conflicts of interest	Court	10/31/2003	$432,750,000
Enron	Earnings manipulation	Court	7/30/2003	$422,995,012
Invesco/AIM	Market timing trading in mutual funds	SEC	10/8/2004	$375,840,004
Bank of America	Market timing trading and late trading in mutual funds	SEC	2/9/2005	$375,000,000
Fannie Mae	Fraudulent accounting	Court	8/9/2006	$350,000,000
Time Warner	Overstating online revenue and number of Internet subscribers	Court	3/29/2005	$300,000,000
Qwest	Overstating income	Court	6/22/2004	$253,606,432
Alliance	Market timing trading in mutual funds	SEC	4/28/2005	$250,850,003

Source: SEC.

Table 3. Fair Fund Orders and Distributions, as of June 2007

	SEC-overseen Fair Funds	Court-overseen Fair Funds	All Fair Funds
Number of plans	46	69	115
Total amount ordered (in thousands)	$3,934,371	$4,512,860	$8,447,231
Total amount distributed (in thousands)	$644,450	$1,122,351	$1,766,802
Percent distributed	16.4	24.9	20.9

Source: GAO analysis of SEC data.

Fair Fund Distribution Delays Have Been Attributed to Difficulties in Identification of Harmed Investors, the Complexity of Certain Cases, and Tax Issues

According to Enforcement officials and consultants who work on Fair Funds, a key reason for the slow distribution of Fair Funds has been the difficulty of identifying harmed investors in certain cases. Unlike typical securities class action lawsuits, Fair Funds may not rely on a claims-based process in which injured parties identify themselves by filing claims with trustees or other administrators. For example, in Fair Fund cases involving mutual fund market timing abuses, which account for many funds ordered into Fair Fund plans, Enforcement attorneys and plan administrators have assumed responsibility for identifying harmed investors. This step was taken because with the large number of affected investors and the nature of market timing violations, many such investors may not even have been aware that their accounts experienced losses.[31] One Fair Fund plan consultant said that many harmed investors already had redeemed their shares in the affected mutual fund companies in prior years. Tracking down such former customers can be challenging because they may have changed their addresses several times, the consultant said. Several consultants and Enforcement officials also said that tracking down customers can be hard because securities brokers, through whom individuals may purchase mutual funds, may maintain customer account information rather than the mutual fund company itself.[32] As a result, a Fair Fund administrator might need to contact and obtain the cooperation of relevant broker-dealers to obtain customer account information and make related distributions.

The complexity of some cases can also impede the timely distribution of Fair Funds. For example, in mutual fund market timing cases, sophisticated analysis might be required to first identify trades that benefited from improper activity and then to calculate profits earned from those transactions and associated losses to investors, which may be spread across many such customers.[33] According to a Fair Fund plan consultant and Enforcement officials, another significant challenge to the Fair Fund distribution involves retirement plans and the Employee Retirement Income Security Act of 1974 (ERISA), the federal law setting minimum standards for pension plans in private industry.[34] Retirement plans hold assets on behalf of their beneficiaries, and it is not unusual for those assets to be invested with entities that become subject to Fair Fund enforcement actions. Thus, ERISA-covered retirement plans will be entitled to Fair Fund proceeds by virtue of such investments. But depending on circumstances, a Fair Fund distribution consultant may need to make determinations on a variety of complex issues before funds can be distributed, such as determining when such distributions become plan assets under ERISA.[35] One Fair Fund consultant told us he spent a year waiting for Department of Labor clarification of relevant ERISA issues.

Finally, determining the tax treatment of funds may also slow the distribution process. According to Fair Fund consultants, tax information must accompany Fair Fund distributions to investors so that recipients have some idea of how to treat their payments for tax purposes. Consultants and Enforcement officials told us that determining appropriate tax treatment has been time-consuming as they had no precedents upon which to draw. Depending on circumstances, an investor's recovery of disgorged profits can constitute ordinary income or a capital gain—which can be taxed at different rates—or not represent taxable income at all. SEC ultimately hired a consulting firm to handle tax issues. One Fair Fund consultant told us that obtaining tax guidance from the Internal Revenue Service delayed the plan's distribution by about 1 year.

Enforcement's Approach to Managing Fair Funds May Also Have Slowed Distributions, and SEC Has Not Defined Responsibilities of a New Office to Administer the Program

In addition to the factors discussed above, Enforcement's largely decentralized approach to managing the process may also have contributed to delays in the distribution of Fair Funds. Currently, Enforcement staff attorneys in either SEC's home office or 1 of its 11 regional offices who are pursuing individual enforcement cases take a lead role in the Fair Funds process, overseeing much of the work necessary to establish and maintain a fund. This includes supervising cases directly, overseeing consultants who design or administer distribution plans, and advising or petitioning courts presiding over Fair Fund plans. Enforcement officials said that the approach made sense from a Fair Fund administration standpoint because division attorneys have substantial knowledge of the regulated entity involved and the relevant enforcement action. Enforcement officials also said that senior officials in the home office have always played an important role in the oversight of the program. Their responsibilities have included providing guidance on selecting consultants, leading information-sharing and problem-solving efforts (for example, leading regular conference calls among fund consultants, parties involved in Fair Fund enforcement actions, their legal counsel, and SEC staff in Enforcement and elsewhere), and reviewing proposed Fair Fund distribution plans and recommending modifications as necessary.

Outside consultants hired to design and implement Fair Fund plans told us that Enforcement staff attorneys assigned to their cases were dedicated and responsive and that the agency appears to be making a good faith effort to implement and oversee the Fair Funds provision. However, they also said that Enforcement's delegated approach has resulted in delays, higher costs, and unnecessary repetition of effort. With different Enforcement staff handling different Fair Fund cases, the consultants said that Enforcement forgoes the opportunity to build institutional expertise and efficiencies. For example, one consultant said that Enforcement's delegated management of the Fair Fund program has resulted in inefficiencies in key administrative aspects of the program, such as the development of standardized means of communicating with investors (for example, form letters) and the mechanics for distributing funds to them. Consequently, the consultant said that the Fair Fund program incurs a substantial amount of unnecessary administrative costs. Further, the consultants generally agreed that it would make sense for SEC to consider centralizing at least some aspects of the administration of Fair Fund plans to improve the efficiency of the distribution process.

While Enforcement officials have cited benefits associated with the current management of the program, both SEC and division officials also acknowledged that it has created challenges. An official within the Chairman's office said that the slow distribution of funds to harmed investors is of significant concern to the agency and that the lack of a centralized management approach has limited the development of standardized policies and controls necessary to facilitate disbursements. Further, Enforcement officials said that while Fair Fund work is important, it can divert investigative attorneys from pursuing other cases. The officials said that the Fair Fund workload on any particular case varies over time, but during peak periods it can consume about 50 to 75 percent of a staff attorney's time. At one SEC regional office, Enforcement officials said that administering the Fair Fund program has resulted in a significant commitment of attorneys' time, especially because the office lost almost 25 percent of its investigative staff due to attrition in the past year or so.

In response to concerns about the slow distribution of Fair Fund proceeds to harmed investors, SEC's Chairman took two actions in 2007. First, he established an internal agency committee to examine the program's operation. The committee—which includes representatives from Enforcement, General Counsel, the Office of the Secretary, and the Office of the Executive Director—is assessing lessons learned in program implementation, the agency's selection of consultants to administer the plans, and SEC's policies and procedures for managing the program. An Enforcement official said that the committee is expected to complete its analysis by September 30, 2007. Second, in March 2007, the Chairman announced plans to create a centralized Fair Fund office.[36] The Chairman stated that the purpose of the new office is to develop consistent fund distribution policies and dedicate full-time trained staff to ensure the prompt return of funds to harmed investors.

While creating a central office within SEC could facilitate the distribution of Fair Funds, it was not yet possible to assess the planned office's potential impact at the time of our review. For example, SEC had not announced which SEC unit the office would report to, although one official said that the office probably would be located within Enforcement. Further, SEC had not staffed the new Fair Fund office and had not established the roles and responsibilities of the new office or written relevant policies and procedures.[37] For example, SEC had not determined the extent to which the new office might assume complete responsibility for managing at least some Fair Fund plans, although it is expected the office will continue to provide support to division attorneys who currently manage such plans. Until such issues are resolved, the new office's potential efficiency in more quickly distributing Fair Fund proceeds to harmed investors will not be realized.

Enforcement Does Not Yet Collect Key Data Necessary to Effectively Oversee the Fair Funds Program

Enforcement does not collect key data, as we recommended in 2005, to aid in division oversight of the Fair Fund program.[38] In particular, Enforcement does not systematically collect data on administrative expenses for all ongoing Fair Fund plans. These costs range from fees and expenses that Fair Fund administrators and consultants charged to the costs of identifying harmed investors and sending checks to them. Approximately two-thirds of individual Fair Funds pay for administrative costs from fund proceeds, so that the greater the administrative expenses, the less money is available for distribution to harmed investors, according to our analysis of SEC Fair Fund information.[39] However, without data on administrative expenses charged, Enforcement cannot judge the reasonableness of such fees and take actions as necessary to minimize them.

Enforcement officials generally attributed SEC's inability to implement our 2005 recommendation to changes in priorities for the development of the agency's information systems. After we issued our 2005 report, Enforcement officials said they began working to modify the CATS system so that it could better track Fair Fund administrative expenses and other data. However, SEC ultimately decided to accelerate the development of a new financial management system for the division, called Phoenix. SEC and Enforcement officials said that the agency implemented the first phase of Phoenix in February 2007. The first phase includes limited information relevant to the Fair Fund program (the amount of money ordered in penalties and disgorgement and the amount paid to the agency), but it does not include data on such items as fund administrative expenses. Enforcement officials told us that a second phase of the Phoenix system will contain additional features for more complete management and monitoring of Fair Fund activity, consistent with our 2005 recommendation. According to Enforcement officials, Phoenix II has been funded and is expected to be in place in 2008. Until Phoenix II is implemented and tested, Enforcement officials will continue to lack information necessary for effective Fair Fund management and oversight.

We also note that in the meantime, Enforcement has not leveraged reports that could enhance the division's understanding of Fair Fund expenses, including administrative expenses. SEC rules generally require that final accounting reports be prepared when SEC-overseen Fair Funds are fully distributed and officially closed.[40] We reviewed four such reports and found that three of them were inconsistent in data reported and did not include comprehensive accounting information. For example, two of the accounting reports did not include complete data on the expenses incurred to administer the Fair Fund plan. Further, senior Enforcement officials said that the division could improve its analysis of information contained in the reports. As a result, Enforcement cannot evaluate the reasonableness of administrative expenses for individual Fair Fund plans or potentially gain a broader understanding of the reasonableness of such expenses among a variety of plans.

ENFORCEMENT COORDINATES ITS INVESTIGATIVE ACTIVITIES INTERNALLY AND WITH OTHER AGENCIES AND IS IN THE PROCESS OF DOCUMENTING CRIMINAL REFERRALS

Enforcement has established a variety of processes to coordinate its investigative and law enforcement activities with other SEC offices. Further, Enforcement has established processes to coordinate its investigative activities with other law enforcement agencies, including Justice. However, Enforcement and SEC have not yet implemented our 2005 recommendation that they document referrals of potential criminal activity to other agencies, although plans to do so have been established as part of the division's new investigation management information system (the Hub).[41] Until Enforcement completes this process, its capacity to effectively manage the referral process is limited.

Enforcement Coordinates with Other SEC Units through Regular Meetings and a Referral Process

Enforcement officials said that they hold a variety of meetings periodically to coordinate investigative and other activities within SEC. As discussed previously, senior Enforcement officials said they meet regularly with the SEC Chairman and commissioners to establish investigative priorities. According to the Director of Enforcement, she meets weekly with the heads of other SEC divisions, and other senior division officials said that they meet periodically with their counterparts in other agency units. Enforcement officials cited their coordination with other SEC units on investigations of the backdating of stock options as an example of the agency's successful collaborative efforts.[42] One Enforcement official said that division staff worked closely with the Office of Economic Analysis to analyze financial data and trends related to options backdating, which allowed them to identify patterns used to target companies for further investigation. This official said that Enforcement also collaborated with the Office of the Chief Accountant, the Division of Corporation Finance, and the Office of the General Counsel throughout this effort.

Enforcement officials also said that coordinating their activities with OCIE is particularly important and that the division places a high value on referrals it receives from OCIE regarding potentially illegal conduct. Enforcement officials said that because OCIE staff regularly examine broker-dealers, investment advisers, and other registered entities, they have a broad perspective on compliance with securities laws and regulations. Enforcement officials in SEC's Philadelphia regional office estimated that about 30 or 35 percent of the enforcement actions the Philadelphia office initiates are based on referrals from OCIE staff. They cited one notable recent insider trading case—involving broker-dealer Friedman, Billings, Ramsey & Co., Inc., and which was among the first cases of its kind since the 1980s—as stemming from a referral from an OCIE examination.[43]

However, other Enforcement officials said that historically they have had some concerns about limitations in information contained in OCIE referrals. These concerns centered on how clearly OCIE identifies potentially improper conduct in its referrals and how much evidence it provides in support of such matters. As a result, in November 2006, OCIE and Enforcement instituted a process that would provide a more formal review of the nature and quality of OCIE referrals. According to OCIE and Enforcement officials, the new procedures expand and formalize a preexisting committee process for reviewing OCIE referrals to Enforcement and communicating the ultimate outcome of those referrals to OCIE. The officials said the revised procedures were instituted to (1) help identify the types of OCIE referrals that were likely (or not) to result in enforcement actions and (2) provide better information to OCIE on the ultimate disposition of its referrals.

Enforcement officials noted that the division receives many more referrals from OCIE than from any other SEC division or office; therefore, developing a formal committee and tracking process for other internal referrals has not been viewed as warranted. SEC also receives referrals from self-regulatory organizations (SRO)—such as what is now the Financial Industry Regulatory Authority (FINRA)—often involving allegations of insider trading, which are received by Enforcement's Office of Market Surveillance (OMS).[44] In a forthcoming report, we assess OMS's and Enforcement's processes for reviewing and prioritizing these SRO referrals.

Enforcement Coordinates with Law Enforcement and Other Regulators and Plans to Document Criminal Referrals in New Information Management System

Enforcement officials also said that division staff have established processes to coordinate their investigative activities and working relationships with other law enforcement and regulatory agencies. For example, Enforcement officials in

SEC's regional offices said they have established effective working relationships with U.S. attorney offices to prosecute alleged criminal violations of the securities laws. In our 2005 report, we discussed how Enforcement worked with Justice and state attorneys general to prosecute investment advisers that allegedly violated criminal statutes related to market timing and late trading.[45] In some cases, Enforcement details investigative attorneys to Justice to assist in the criminal prosecution of alleged securities law violators. Other outside organizations with which SEC and Enforcement coordinate investigative activities include the Federal Bureau of Investigation, federal banking regulators, the Commodity Futures Trading Commission, state securities regulators, and local police. Enforcement also participates in interagency investigative task forces, such as the Corporate Fraud Task Force, the Bank Fraud Enforcement Working Group, and the Securities and Commodities Fraud Working Group.[46]

Additionally, in March 2007, Enforcement held its annual conference at SEC's Washington headquarters on securities law enforcement, which federal and state regulators and law enforcement personnel attended. Topics covered included coordination of SEC investigations with criminal law enforcement agencies and advice on trying a securities case. SEC also conducted sessions on market manipulation, insider trading, financial fraud, stock options backdating, and executive compensation. In September 2007, Enforcement will join other SEC units in hosting the Commission's second Seniors Summit, at which SEC, other regulators, and law enforcement agencies will discuss how to work together to address the growing problem of fraud targeting the nation's senior citizens.

Although Enforcement officials say they are planning to do so, they have not yet fully implemented our 2005 recommendation to document Enforcement's referrals of potential criminal matters—and the reasons for making them—to other law enforcement agencies.[47] As discussed in that report, SEC has established a policy under which Enforcement attorneys may make referrals on an informal basis to Justice and other agencies with authority to prosecute criminal violations. That is, Enforcement attorneys may alert other agencies to potential criminal activity through phone calls or meetings, and such referrals need not be formally approved by the division or the Commission. We noted that such an informal referral process may have benefits, such as fostering effective working relationships between SEC and other agencies, but also found that Enforcement did not require staff to document such referrals. Appropriate documentation of decision-making is an important management tool. Without such documentation, Enforcement and the Commission cannot readily determine whether staff make appropriate and prompt referrals. Also, the division does not have an institutional record of the types of cases that have been referred over the years. However,

Enforcement officials told us that the forthcoming Hub system will include data fields that indicate when informal referrals of potential criminal activity have been made.

CONCLUSIONS

In recent years, SEC's Enforcement division and investigative attorneys have initiated a variety of high-profile enforcement actions that resulted in record fines and other civil penalties for alleged serious securities violations and contributed to criminal convictions for the most egregious offenses. While Enforcement has demonstrated considerable success in carrying out its law enforcement mission, some significant limitations in the division's management processes and information systems have hampered its capacity to operate at maximum effectiveness and use limited resources efficiently. One key reason for these limitations appears to have been Enforcement's management approach, which emphasized a broad delegation of key functions with limited centralized management review and oversight, particularly in the approval and review of new investigations and the administration of the Fair Fund program. Delegation of authority is an important management principle that can foster creativity at the local level and, in the case of Enforcement, likely had some benefits for the investigative process and the administration of the Fair Fund program. However, without well-defined management processes to exercise some control over delegated functions, inefficient program implementation and resource allocation can also occur.

Officials from Enforcement and the Offices of the Chairman and Executive Director have recognized limitations in the division's operations and taken important steps to establish more centralized oversight procedures. In particular, they have centralized the review and approval of new investigations, moved forward to upgrade or replace information systems key to division operations and management, and announced the creation of a new Fair Fund office. However, as described below, these plans require additional actions to fully address identified limitations and maximize the division's operational effectiveness.

- Enforcement has not developed written procedures and criteria for reviewing and approving new investigations. Establishing such guidance would help focus the review of investigations and reinforce the consistency of reviews, as intended by the centralization of this function, and assist in communicating the new policies to all current and new staff. Further, developing written procedures and criteria would establish a transparent and agreed-upon standard for the review and approval of new investigations and thereby facilitate the Commission's ability to oversee and evaluate the division's operations and resource allocation.

- Enforcement has not developed written controls to help ensure the timely and consistent entry of investigative data in the Hub information system, which could increase the risk of misleading or inaccurate management reports being generated by the system. Without written guidance and the establishment of independent and regular reviews of the accuracy of Hub data by division officials, Enforcement is not well positioned to help ensure that it is receiving reliable program information. Further, the lack of guidance and controls may limit the new system's capacity to better manage the investigation process.

- Enforcement's potentially large backlog of investigations for which closing memoranda and other required administrative procedures have not been completed requires division management's attention. We recognize that clearing this potentially large backlog could pose challenges to Enforcement given the resource commitment that would be required to do so. Nevertheless, leaving such investigations open indefinitely continues to compromise management's ability to effectively manage its ongoing portfolio of cases. Moreover, it has potentially negative consequences for individuals and companies that are no longer under investigation.

- SEC has not yet staffed or defined the roles and responsibilities of the new office that is being established to administer the Fair Fund program. Therefore, it is not possible to determine the extent to which the office may better facilitate the distribution of funds to investors harmed by securities frauds and other violations. While Enforcement awaits the development and implementation of a new information system that would collect comprehensive information on Fair Fund expenses for ongoing plans (for example, administrative expenses), the division has not taken other steps that would, in the meantime, allow it to develop a better perspective on the reasonableness of such expenses. That is, Enforcement has not ensured the consistency of information contained in reports on completed Fair Fund plans or sufficiently analyzed such reports, compromising its capacity to monitor the program.

Given SEC and Enforcement's critical law enforcement mission, it is important that senior officials ensure that weaknesses in their planned improvements be addressed and implementation monitored. Without a full resolution of existing limitations, a significant opportunity to further enhance the division's effectiveness may be missed.

APPENDIX I: SCOPE AND METHODOLOGY

To address our first objective—evaluating the Securities and Exchange Commission (SEC) Division of Enforcement's (Enforcement) internal processes and information systems for planning, tracking, and closing investigations and planned changes to these processes and systems—we reviewed relevant SEC and Enforcement documentation and data, including the agency's strategic plan, annual performance reports, performance measurement data, investigation and enforcement action data from the Case Activity Tracking System (CATS), and Enforcement personnel data. We also reviewed guidance on Enforcement's intranet—EnforceNet—to determine internal procedures for conducting and managing the investigation process and obtained documentation and attended demonstrations for the first phase of Enforcement's new planned successor system for CATS (the Hub) and for the base model system for the Hub (M&M).[1] We also reviewed prior GAO reports on SEC and Enforcement processes and information technology systems, as well as federal internal control standards.[2] Further, we interviewed the SEC Chairman, two commissioners, officials from SEC's Offices of the Executive Director and General Counsel, and Enforcement officials in Washington and the agency's New York, Boston, and Philadelphia regional offices.

To address our second objective—evaluating the implementation of SEC's Fair Fund responsibilities—we reviewed a 2005 GAO report that discussed SEC's Fair Fund process, as well as relevant legislation.[3] We also obtained and analyzed summary Fair Fund statistics and documentation and data on individual funds, as provided by Enforcement.[4] However, Enforcement did not provide data on 24 Fair Fund plans that were identified in our 2005 report. Among the reasons Enforcement officials cited for the omissions were that some of the 24 funds had been fully distributed and thus were not included in an information system established in 2006 that was designed to track only ongoing plans. However, these 24 Fair Fund plans are generally smaller (accounting for about $118 million or 1 percent of total Fair Funds), and their exclusion does not change our overall conclusion that distributions have been limited.

In addition to Fair Fund data, we reviewed SEC guidance on Fair Funds, including rules on distribution plans, tax treatment, selection of consultants, and distribution procedures. We also reviewed Fair Fund guidance from the U.S. Department of Labor. In addition to discussing the Fair Fund program with relevant SEC and Enforcement officials, we interviewed six consultants hired to design and implement Fair Fund plans, attorneys, consumer advocates, an academic expert, and a representative of a trade group for a retirement plan service provider trade group.

To address the third objective—evaluating Enforcement's efforts to coordinate investigative activities with other SEC divisions and federal and state law enforcement agencies—we reviewed previous GAO reports on mutual fund trading abuses and Enforcement's coordination efforts with law enforcement.[5] We also reviewed relevant SEC documentation, including internal referral policies, and guidance regarding coordination between Enforcement and outside law enforcement authorities. We also attended SEC's annual securities coordination conference held in Washington in March 2007, which was attended primarily by federal and state regulators and law enforcement personnel. Further, we discussed Enforcement's coordination efforts with relevant SEC and division officials.

We conducted our work in Washington, D.C.; Boston, Massachusetts; Philadelphia, Pennsylvania; and New York, New York, between November 2006 and July 2007 in accordance with generally accepted government auditing standards.

APPENDIX II: SELECTED DIVISION OF ENFORCEMENT INVESTIGATION AND PERSONNEL DATA

We collected data from the Securities and Exchange Commission (SEC) on the Division of Enforcement's (Enforcement) investigative caseload and other personnel information (see tables 4-6 below). As shown in table 4, the ratio of ongoing Enforcement investigations to staff attorneys increased substantially from about five investigations per attorney in 2002 to eight per attorney in 2006, according to SEC data. However, these SEC data should be interpreted with caution and may significantly overestimate the number of investigations per Enforcement attorney. The reported number of investigations includes all open investigations at the end of each year, even investigations that have been open for many years. As discussed in this report, Enforcement has not promptly closed many investigations that have not resulted in enforcement actions and are likely no longer being actively pursued. Accordingly, we requested that SEC provide data on the number of ongoing investigations in Enforcement that as of year-end 2006 had been initiated within the previous 2 years. When pending investigations that were more than 2 years old are excluded, the investigation-to-staff-attorney ratio drops to 2.54.[1] While this ratio may provide a more accurate assessment of Enforcement attorneys' active workloads, individual investigations more than 2 years old could continue to be actively pursued while some individual investigations less than 2 years old may no longer be actively pursued. Enforcement officials estimated that staff attorneys generally can be working on from 3 to 5 investigations at one time, including administering individual Fair Fund plans.

Table 4. Ratio of Open Investigations to Staff Attorneys

2002	2003	2004	2005	2006
5.07	5.43	6.57	7.20	8.06

Source: GAO analysis of SEC data.

Note: Open investigations refer to the number of investigations pending as of the end of each fiscal year according to SEC's annual reports. Staff attorneys do not include trial attorneys and accountants.

Table 5 shows that the ratio of Enforcement investigative attorneys to paralegals, who provide support to the investigative process, generally declined from 2003 to 2006. Our review of SEC data indicates that the number of Enforcement paralegals increased substantially from 2003 to 2005 (from 58 to 98, or 69 percent) and remained stable in 2006 at 94 (a decline of just over 4 percent). While the number of Enforcement staff and supervisory attorneys also increased from 2003 to 2005 (from 596 to 740, or 24 percent), the rate of increase was not nearly as high as for paralegals. In addition, the number of Enforcement investigative attorneys declined from 740 in 2005 to 684 in 2006, or 8 percent. The relatively slower pace of attorney hiring from 2003 to 2005 and relatively higher rate of attrition in 2006 helps explain why the ratio of attorneys to paralegals has declined in recent years. Other SEC data that we reviewed also indicated a decline in the ratio of investigative attorneys to various types of administrative support staff, such as research specialists, during this period.

Table 5. Ratio of All Investigative (Staff and Supervisory) Attorneys to Paralegals

2002	2003	2004	2005	2006
10.28	11.88	9.00	7.55	7.28

Source: GAO analysis of SEC data.

Note: All investigative attorneys include staff and supervisory attorneys in the Enforcement division. Supervisory attorneys refer to positions that SEC personnel data label as "(supervisory) general attorneys." This does not include (supervisory) trial attorneys and (supervisory) accountants. SEC personnel data label paralegals as "paralegal specialists."

Table 6 shows that the ratio of investigative staff attorneys to supervisory attorneys has remained relatively constant. Supervisory attorneys are branch chiefs and assistant directors and do not include attorneys at the associate director level and above.

Table 6. Ratio of Staff Attorneys to Supervisory Attorneys

2002	2003	2004	2005	2006
3.20	3.59	3.50	3.30	3.02

Source: GAO analysis of SEC data.

REFERENCES

[1]   Disgorgement deprives securities law violators of ill-gotten gains linked to their wrongdoing.

[2]   15 U.S.C. § 7246.

[3]   See GAO, Mutual Fund Trading Abuses: SEC Consistently Applied Procedures in Setting Penalties, but Could Strengthen Internal Controls, GAO-05-385 (Washington, D.C.: May 16, 2005). This report generally addressed SEC enforcement actions pertaining to market timing and late trading violations. Market timing typically involves the frequent buying and selling of mutual fund shares by sophisticated investors who seek opportunities to make profits on the difference in prices between overseas and U.S. markets. Late trading is illegal and occurs when investors place orders to buy or sell mutual fund shares after the mutual fund has calculated the price of its shares but still receive that day's fund share price. As of February 2005, Enforcement had initiated 24 enforcement

actions that resulted in fines of almost $2 billion against mutual fund companies and officers for market timing and late trading violations.

[4]   Pub. L. No. 107-204 § 601, 116 Stat. 745 (July 30, 2002). Amelia Gruber, "SEC Urged to Flesh Out Performance Goals," *GovernmentExecutive.Com* (Washington, D.C.: Sept. 28, 2004).

[5]   GAO, SEC Operations: Oversight of Mutual Fund Industry Presents Management Challenges, GAO-04-584T (Washington, D.C.: Apr. 20, 2004).

[6]   GAO-04-584T.

[7]   Section 308(a) of the Sarbanes-Oxley Act of 2002, entitled "Fair Fund for Investors," allowed SEC to combine civil monetary penalties and disgorgement amounts collected in enforcement cases to establish funds for the benefit of victims (investors) of securities law violations. 15 U.S.C. § 7246. See also GAO, *SEC and CFTC Penalties: Progress Made in Collection Efforts, but Greater SEC Management Attention Is Needed*, GAO-05-670 (Washington, D.C.: August 2005).

[8]   CATS contains information on ongoing investigations and enforcement actions, such as the general nature of the potential violation (for example, insider trading) and the date an investigation was opened.

[9]   A hedge fund is generally an entity that holds a pool of securities and other assets, is not required to register its securities offerings, and is excluded from the definition of an investment company.

[10]   Due to deficiencies in CATS, Enforcement cannot readily provide data on the number of ongoing investigations that have not resulted in enforcement actions.

[11]   These 115 Fair Funds are tracked in Enforcement's distribution management system. CATS tracks all Fair Fund distributions that have occurred but by defendant, not by fund. When a Fair Fund is created through an SEC administrative action, SEC oversees the case directly. When a fund is created through court action, SEC is a party to the court proceeding, but the court retains ultimate authority to supervise the plan. In either an administrative or court proceeding, an individual or company is ordered or agrees to pay an amount of money into a Fair Fund plan. The Fair Fund data discussed in this report do not include 24 cases that SEC had previously identified as Fair Funds. These 24 cases generally are smaller (accounting for about $118 million or 1 percent of total Fair Funds), and their exclusion does not change the overall conclusion that distributions have been limited. Enforcement did not include some of the 24 plans because they were fully distributed to harmed investors and the

division's information system only tracks Fair Fund plans that were ongoing at the time this system was established in 2006.

[12] GAO-05-670.

[13] Section 201.1105(f) of title 17 of the Code of Federal Regulations, SEC's Rules of Practice regarding Fair Fund and Disgorgement Plans, generally provides, *inter alia,* that "a final accounting shall be submitted for approval of the Commission or hearing officer prior to discharge of the administrator and cancellation of the administrator's bond, if any." SEC also seeks to track activity and expenses of court-overseen Fair Funds.

[14] GAO-05-385. In this report, "referrals" are Enforcement's interactions and consultations with law enforcement agencies on specific cases rather than the formal referral process mentioned in our 2005 report, which Enforcement no longer uses.

[15] SEC used to utilize regional offices to oversee district offices, but as of March 2007, all 11 offices were designated regional offices.

[16] Both deputy directors used to be located in the home office, but in April 2007, one relocated to the New York regional office.

[17] Regional offices comprise positions that (1) belong exclusively to Enforcement, (2) are shared by Enforcement and other teams, and (3) do not belong to Enforcement at all. For the purposes of computing the total Enforcement staff numbers, we counted only those positions that belonged exclusively to Enforcement. Pub. L. No. 107-204, tit. VI, § 601, 116 Stat. 745 (July 30, 2002); 15 U.S.C. § 78kk.

[18] The investigative attorney numbers include Enforcement staff with position titles of general attorney or supervisory general attorney, which do not include attorneys above an assistant director level. Examples of position titles not included in these numbers, but included in the numbers of total Enforcement staff, are case management specialist, law clerk, legal technician, paralegal specialist, research specialist, secretary, staff accountant, and trial attorney.

[19] After being open for more than 60 days, an MUI automatically becomes an investigation.

[20] GAO-05-385.

[21] 15 U.S.C. § 7246.

[22] The Chairman stated that millions of individuals are expected to retire in the coming decade, meaning they will need to actively manage their investment accounts, and many individuals and companies may seek to take advantage of this increased investment activity and defraud them of their savings. Testimony of SEC Chairman Cox before the U.S. House

Subcommittee on Financial Services and General Government, Committee on Appropriations, 110th Congress, 1st session, March 27, 2007.

[23] An Enforcement official told us that the 40 percent limit was established based on analysis of 20 years of enforcement action data.

[24] This process does not apply to MUIs that are opened based on referrals from SROs. Referrals from SROs are sent directly to Enforcement's Office of Market Surveillance in the home office, which then reviews the referral, decides whether or not to open a MUI, and, if one is opened, sends the MUI to the appropriate office for action.

[25] See GAO, *Standards for Internal Control in the Federal Government*, GAO/AIMD-00.21.3.1 (Washington, D.C.: November 1999) and Office of Management and Budget, *Management's Responsibility for Internal Control* , OMB Circular No. A-123 Revised, Appendix A, "Internal Control Over Financial Reporting" (Washington, D.C.: December 2004).

[26] GAO/AIMD-00.21.3.1.

[27] Enforcement officials said that ongoing investigation data maintained in CATS will be electronically transferred to the Hub when the system is implemented. However, division attorneys will be responsible for entering relevant data into the Hub for ongoing investigations that are not maintained in CATS, such as detailed information on the type of investigation (e.g., whether it is a hedge fund investigation).

[28] In addition to directing federal agencies to establish written controls, GAO/AIMD-00.21.3.1 and OMB Circular A-123 call for the establishment of controls to ensure the reliability of data maintained in information systems.

[29] We requested that SEC provide data on the number of investigations open for 2 or more years that have outstanding enforcement actions. To respond to this request, an Enforcement official said that OIT would have had to spend a great deal of time creating complex programs. Due to other demands on OIT's resources and our ability to obtain related data from an SEC regional office, we decided not to request that OIT provide this information.

[30] GAO-05-670. The Fair Fund data do not include 24 cases that SEC had previously identified as Fair Funds. These 24 cases generally are smaller (accounting for about $118 million or 1 percent of total Fair Funds), and their exclusion does not change the overall conclusion that distributions have been limited. Enforcement did not include some of the 24 plans because they were fully distributed to harmed investors and the division's

information system tracks only Fair Fund plans that were ongoing at the time this system was established in 2006.

[31]   Market timing losses generally were distributed across many individual mutual fund customers. The losses were often small and investors may not even have realized that their account balances were experiencing dilution for extended periods. They also may have redeemed their shares in the mutual fund company while market timing violations were occurring.

[32]   Broker-dealers may maintain such customer account information on an aggregated basis in what are known as omnibus accounts, and the mutual fund would not have direct access to this information.

[33]   Market timing is said to dilute the value of mutual fund shares, as a market timer buys, sells, or exchanges shares rapidly and repeatedly to take advantage of favorable prices. In addition, market timing increases transaction costs for mutual funds. To take account of investor losses due to dilution, a Fair Fund distribution plan might attempt to estimate, on a daily basis, the extent to which a fund's net asset value (analogous to share price) would have been more or less than the actual net asset value had market timing not occurred. The difference, where positive, is the estimate of dilution and harm to investors. The sum of daily increments (both positive and negative) represents aggregate harm to a fund's shareholders over the period in which market timing occurred.

[34]   Pub. L. No. 93-406, 88 Stat. 829 (Sept. 2, 1974).

[35]   In some cases, retirement plans will be shareholders of record and receive Fair Fund distributions directly. In other cases, an intermediary—such as a broker-dealer, underwriter, or record-keeper—will be the shareholder of record, and retirement plans will receive Fair Fund distributions based on their interest in an account operated by the intermediary.

[36]   See testimony of SEC Chairman Cox before the U.S. House Subcommittee on Financial Services and General Government, Committee on Appropriations, 110th Congress, 1st session, March 27, 2007.

[37]   GAO/AIMD-00.21.3.1 and OMB Circular A-123.

[38]   GAO-05-670.

[39]   According to information that SEC provided us, 81 of 115 Fair Funds, or 70 percent, have provisions whereby fund proceeds are used to pay administrative expenses. In the remaining 34 cases, the individual or entity sued in the relevant enforcement action, such as a mutual fund company, pay Fair Fund expenses.

[40]   Section 201.1105(f) of title 17 of the Code of Federal Regulations. SEC's Rules of Practice regarding Fair Fund and Disgorgement Plans generally provides, *inter alia,* that "a final accounting shall be submitted for approval of the Commission or hearing officer prior to discharge of the administrator and cancellation of the administrator's bond, if any." SEC also seeks to track activity of court-overseen Fair Funds.

[41]   GAO-05-385.

[42]   In a typical case, companies misrepresent the date on which stock options were granted (using a date on which the price was lower). When the holders exercise their options, they can realize larger gains because their exercise prices are based on the lower, misrepresented grant date; the company meanwhile doesn't report the larger gains as greater compensation. The practice violates SEC's disclosure and accounting rules, and tax laws.

[43]   *SEC v. Friedman, Billings, Ramsey & Co., Inc.* No. 06-cv-02160 (D.D.C. 2006), SEC Litigation Release No. 19950 (December 20, 2006). According to SEC Release No. 19950, Friedman, Billings, Ramsey, & Co., Inc. agreed to settle the matter without admitting to or denying the allegations.

[44]   FINRA was created in July 2007 through the consolidation of NASD (formerly an SRO) and the member regulation, enforcement, and arbitration functions of the New York Stock Exchange; it is now the largest nongovernmental regulator for all securities firms doing business in the U.S.

[45]   GAO-05-385.

[46]   The Corporate Fraud Task Force includes a Department of Justice group that focuses on enhancing criminal enforcement internally and an interagency group that focuses on cooperation and joint federal regulatory and enforcement efforts. The Bank Fraud Enforcement Working Group promotes coordination and communication among financial institution regulators and federal law enforcement authorities. The Securities and Commodities Fraud Working Group provides a forum for federal law enforcement authorities to exchange information with securities and commodities regulators, securities SROs, and the Public Company Accounting Oversight Board.

[47]   GAO-05-385.

Appendix I

[1]   In 2004, the Boston regional office tasked one of its staff members and an outside consultant to design a new case tracking system for the office. This new database was implemented in 2005 and also piloted in the Los Angeles and Chicago regional offices. When beginning to develop the Hub in 2006, SEC decided to use Boston's system as the base model for the Hub because of the amount of user input M&M already had received.

[2]   GAO, Mutual Fund Trading Abuses: SEC Consistently Applied Procedures in Setting Penalties, but Could Strengthen Internal Controls, GAO-05-385 (Washington, D.C.: May 16, 2005) and Standards for Internal Control in the Federal Government, GAO/AIMD-00-21.3.1 (Washington, D.C.: Nov. 1999).

[3]   GAO, SEC and CFTC Penalties: Continued Progress Made in Collection Efforts, but Greater SEC Management Attention Is Needed, GAO-05-670 (Washington, D.C.: Aug. 31, 2005).

[4]   Individual Fair Fund data obtained included information on the type of enforcement action that produced the Fair Fund, type of adjudication, amounts ordered placed into a Fair Fund and amounts distributed, dates of issuance of Fair Fund orders, and whether parties involved were responsible for paying the expenses of their respective Fair Fund plan.

[5]   GAO-05-385 and GAO, *U.S. Attorneys: Performance-Based Initiatives are Evolving*, GAO-04-422 (Washington, D.C.: May 28, 2004).

Appendix II

[1]   We computed this ratio by dividing the number of investigations with a status of "active" in Enforcement's information system (CATS) that had been open for less than 2 years as of December 31, 2006—1,305—by the number of "staff attorneys" (nonsupervisory investigative attorney positions that SEC personnel data label as "general attorneys") as of September 31, 2006—514.

Chapter 3

STEPS BEING TAKEN TO MAKE EXAMINATION PROGRAM MORE RISK-BASED AND TRANSPARENT*

ABBREVIATIONS

CCO	Chief Compliance Officer
NASD	National Association of Securities Dealers
NYSE	New York Stock Exchange
OCIE	Office of Compliance Inspections and Examinations
ORA	Office of Risk Assessment
OIT	Office of Information Technology
RADAR	Risk Assessment Database for Analysis and Reporting
SEC	Securities and Exchange Commission
SRO	self-regulatory organizations

---

* Excerpted from GAO Report GAO-07-1053, dated August 2007.

August 14, 2007
Congressional Requesters:

The authority of the Securities and Exchange Commission (SEC) to conduct inspections and examinations of certain participants in the securities industry is one of its most important tools in detecting fraud and violations of securities laws. SEC exercises this authority through its Office of Compliance Inspections and Examinations (OCIE). In fiscal year 2006, OCIE conducted over 2,600 examinations of investment companies, investment advisers, broker-dealers, and other securities-related firms registered with SEC (registrants).[1]

After widespread unlawful trading practices in the mutual fund industry surfaced in late 2003, OCIE attempted to address concerns about the effectiveness of its ability to detect such practices in its examinations of registrants by revising its examination approach to try to better identify and focus its limited resources on those activities representing the highest risk to investors.[2] To ensure registrants understand and address weaknesses in compliance and violations found during examinations, OCIE has formal exit procedures for examiners to follow when communicating the findings of examinations to registrants. However, some registrants—including investment companies, investment advisers, and broker-dealers—have raised concerns about OCIE staff's not communicating the status and results of examinations. In May 2006, the SEC Chairman testified before the House Financial Services Committee on recent changes to the examination program, which were designed to further increase communication with registrants as well as enhance preexamination planning.[3] These reforms include a new procedure that, among others, requires examiners to contact registrants when an examination extends 120 days beyond the on-site visit and alert them to the status of the examination.

This chapter addresses your interest in OCIE's progress toward more risk-based examinations for registered investment companies and investment advisers, implementation of recent initiatives in the examination program, and efforts to communicate key examination information to registrants and minimize disruption. Specifically, we (1) describe how OCIE revised the examination approach after 2003 for investment companies and investment advisers registered with SEC; (2) discuss OCIE's exit procedures and the frequency with which examiners have followed these procedures when conducting examinations; and (3) describe reforms OCIE implemented since January 2006 to increase communication with registrants and improve the examination program, including how examiners complied with the new 120-day notification requirement.

To address the first objective, we analyzed information obtained through OCIE documents and interviews with OCIE and other SEC officials on OCIE's revised examination approach for investment companies and investment advisers and a new process for identifying risks in the marketplace. We also observed a demonstration of the information-technology application that OCIE uses to conduct its annual risk-assessment process. To address the second objective, we reviewed OCIE's guidance to examiners, interviewed OCIE officials on exit procedures, and reviewed examination data. We selected two random samples of 129 examinations, one from the population of investment company and investment adviser examinations and one from the population of broker-dealer examinations completed during fiscal year 2003 through fiscal year 2006. This process allowed us to project our results to the two respective populations at the 95 percent level of confidence. All estimates in this report have margins of error of plus or minus 8 percent or less. To address the third objective, we reviewed memorandums from OCIE to the Commission and the revised examination brochure, analyzed examination data related to the notification procedure, interviewed officials from OCIE, and obtained the views of various industry participants representing investment companies, investment advisers, and broker-dealers. In determining the frequency with which examiners complied with the new notification procedure, we identified all closed examinations that lasted 120 days or more conducted between July 31, 2006, the day the guidance was implemented, and February 2, 2007, the day OCIE gave us the records. We reviewed all 13 examinations that met these criteria. In conducting our analyses of examination data, we conducted a data reliability assessment of the data OCIE provided us and determined it was reliable for our purposes.

We performed our work in Washington, D.C., between October 2006 and July 2007 in accordance with generally accepted government auditing standards. Appendix I provides a more detailed description of our scope and methodology.

RESULTS IN BRIEF

Since 2003, when SEC and state securities regulators discovered widespread unlawful conduct in mutual fund trading by investment advisers and other service providers, OCIE has revised its approach to examining registered investment companies and investment advisers to try to better identify firms with greater compliance risks as well as emerging industry practices that may have potential compliance issues and to target examination resources accordingly.[4] More specifically, in fiscal year 2005 OCIE shifted its focus from the routine examination of all registered investment companies and advisers, regardless of compliance risks, to the examination of "higher-risk" firms—about 10 percent of the population—once every 3 years. From the remaining 90 percent of the population designated as "lower risk," OCIE examines a small random sample annually. Under OCIE's revised approach, "sweep" examinations, which target specific activities across firms, and "cause" examinations, which target known problems at an individual firm, are also a higher priority. The effectiveness of OCIE's revised approach depends on its ability to accurately assess the level of risk at individual investment advisers; inaccurately categorizing firms as lower-risk could result in harmful practices' going undetected.[5] Since 2002, OCIE has assigned risk ratings to investment advisers after evaluating their compliance controls through routine examination. However, most firms have not yet received this evaluation. To assign risk ratings to unexamined firms, OCIE assesses publicly available information to identify risks inherent in a firm's businesses, such as conflicts of interest. While these variables may indicate areas of high risk, they do not provide any information on the firm's policies or procedures for mitigating these risks. OCIE's analysis of fiscal year 2006 data showed that the accuracy of this methodology for predicting whether firms are higher- or lower-risk has some limitations. OCIE officials said that they are evaluating other potential indicators of compliance risks, such as investment adviser performance, to improve their risk-rating methodology and otherwise aid them in identifying higher-risk firms. Implementation of our prior recommendation to obtain and review documentation associated with the compliance reviews that firms must conduct under SEC rules—a source of information on the effectiveness of their compliance controls—could potentially help OCIE better identify higher-risk firms as part of its risk-assessment methodology.[6]

Our review of investment company, investment adviser, and broker-dealer examinations completed during fiscal year 2003 through fiscal year 2006 found that examiners generally applied OCIE's exit procedures, with the major exceptions occurring during sweep examinations relating to mutual fund trading abuses, instances where OCIE directed examiners to forgo exit procedures. To communicate deficiencies to registrants, OCIE has instituted specific exit procedures that include an exit interview, which examiners use to inform registrants of deficiencies prior to the close of an examination, and a "closure notification" letter, which communicates the outcome of the examination. OCIE guidance allows examiners to refrain from applying these procedures when they refer their findings to the Division of Enforcement (Enforcement) and are asked to forgo the exit interview, the deficiency letter, or both; or when an examination results in no findings, in which case an exit interview is not necessary. OCIE management also has directed examiners to deviate from exit procedures under exigent circumstances, such as during the extensive sweep examinations initiated to address the widespread unlawful trading in mutual funds that surfaced in 2003 and that included inappropriate market timing, among other practices.[7] Our analysis of a sample of investment company and investment adviser examinations completed during fiscal year 2003 through fiscal year 2006 estimated that examiners conducted exit interviews for 79 percent of the examinations completed during this period. They did not conduct interviews in an estimated 12 percent for reasons consistent with their guidance. In the other estimated 9 percent, OCIE directed examiners not to conduct exit interviews because the examinations were part of ongoing sweep examinations related to market timing. We also estimated that examiners sent either a deficiency letter or a "no further action" letter in 87 percent of the examinations. Examiners did not send closure notifications in an estimated 11 percent because the examination was part of the ongoing sweep examinations related to market timing and in 2 percent for other guidance-related reasons. We did not find evidence that examiners sent closure notification letters in the remaining estimated 1 percent, when OCIE guidance indicated they should have been sent.[8] We also analyzed a sample of broker-dealer examinations and estimated that examiners conducted exit interviews and sent closure notifications in 82 percent and 88 percent, respectively, of the total number of examinations completed during the review period and did not conduct these procedures in 11 percent and 9 percent, respectively, of examinations for reasons consistent with OCIE's guidance. However, in an estimated 7 percent of examinations, we did not find evidence of an exit interview when OCIE guidance indicated one was warranted. This estimate includes an estimated 3 percent of cases where OCIE officials told us examiners conducted the interviews but did not document the discussion.

OCIE generally followed its new procedure requiring examiners to inform registrants of the status of examinations extending past 120 days, one of a variety of new initiatives OCIE implemented to improve coordination and communication among examiners, and with other SEC divisions and registrants. Other examples include protocols and tools to help examiners across SEC headquarters and regional offices coordinate their examinations and avoid duplication as well as a hotline for registrants to call with complaints or concerns about the examination program. In reviewing OCIE examination data to determine the extent to which examiners followed the 120-day requirement, we identified 13 closed examinations to which this procedure was applicable. In 12 of the 13, examiners either provided the notification or had a guidance-related reason for not contacting the firm, such as a request by Enforcement. To obtain the views of registrants on OCIE's new initiatives, we contacted various industry participants representing investment companies, investment advisers, and broker-dealers. A number of registrants questioned the effectiveness of the new hotline, as it is located within OCIE's Office of the Chief Counsel and not in another SEC office or division that is independent of OCIE. These registrants said they would hesitate to use the new hotline, thereby limiting its effectiveness as a communication tool.

This chapter contains one recommendation designed to facilitate greater use of OCIE's new examination hotline by relocating it to a division or office that is independent of OCIE. We received comments on a draft of this report from SEC, which are included in appendix II. In its written comments, SEC agreed with our conclusions and noted that in response to our recommendation, OCIE is developing a revised hotline where callers can choose to speak with the Commission's Office of the Inspector General, in addition to staff from OCIE's Office of the Chief Counsel. SEC also provided technical comments on a draft of the report, which were incorporated into the final report, as appropriate.

BACKGROUND

SEC oversees investment companies and investment advisers primarily through OCIE; the Division of Investment Management (Investment Management); and Enforcement. OCIE examines investment companies and investment advisers to evaluate their compliance with federal securities laws, determine if these firms are operating in accordance with disclosures made to investors, and assess the effectiveness of their compliance control systems. Investment Management administers the securities laws affecting investment companies and investment advisers. It reviews the disclosure documents that investment companies registered with SEC are required to file with the agency and engages in other regulatory activities, such as rule making, responding to requests for exemptions from federal securities laws, and providing interpretation of those laws. Enforcement is responsible for investigating and prosecuting violations of securities laws or regulations that are identified through OCIE examinations, referrals from other regulatory organizations, and tips from firm insiders, the public, and other sources.

OCIE conducts routine, sweep, and cause examinations to oversee registered investment companies and investment advisers. Routine examinations are conducted according to a cycle that is based on the registrant's perceived risk. During a routine examination, OCIE assesses a firm's process for assessing and controlling compliance risks. In 2002, OCIE started to use a systematic approach for documenting and assessing the effectiveness of investment advisers' compliance controls.[9] Based on that assessment, examiners assign investment advisers risk-ratings indicating whether they are at higher- or lower- risk for experiencing compliance problems. In a sweep examination, OCIE probes specific activities of a sample of investment companies and investment advisers to identify emerging compliance problems in order that they may be remedied before becoming too severe or systemic. OCIE conducts cause examinations when it has reason to believe something is wrong at a particular registrant. Investment companies and investment advisers can be candidates for cause examinations if they are the subject of investor complaints, tips, or critical news media reports.

SEC regulates broker-dealers in conjunction with National Association of Securities Dealers (NASD) and the New York Stock Exchange (NYSE), among others.[10] NASD and NYSE are self-regulatory organizations (SRO) with statutory responsibilities to regulate their own members. As part of their responsibilities, they conduct examinations of their members to ensure compliance with SRO rules and federal securities laws. OCIE evaluates the quality of NASD and NYSE oversight in enforcing their members' compliance through oversight inspections of the SROs and broker-dealers. SRO oversight inspections review all aspects of an SRO's compliance, examination, and enforcement programs.[11] Through broker-dealer oversight examinations, OCIE re-examines a sample of firms within 6 to 12 months after the SRO completed its examination.[12] In addition to broker-dealer oversight examinations, OCIE also directly assesses broker-dealer compliance with federal securities laws through special and cause examinations. Special examinations include sweep examinations and internal controls risk management examinations of the 20 largest broker-dealer firms. The Division of Market Regulation (Market Regulation) administers the securities laws affecting broker-dealers and engages in related oversight activities such as rule making. Both SEC's Enforcement and the SROs' enforcement divisions are responsible for investigating and disciplining violations of securities laws or regulations by broker-dealers.

OCIE Revised Its Examination Approach to Target Higher-Risk Registrants and Compliance Issues

Since 2003, OCIE changed its examination program for certain registrants including investment companies and investment advisers to try to focus its examination resources on those firms and industry practices with the greatest risk of having compliance problems. In particular, OCIE went from routinely examining registered firms on an established schedule to emphasizing the examination of higher-risk firms. Accurate risk ratings of investment advisers are critical to making this revised approach effective. However, to assign risk ratings to firms that have not had their compliance controls evaluated through routine examinations, OCIE uses proxy indicators for compliance risk that do not incorporate information on the strength of the firm's compliance controls, a limitation that OCIE has recognized. One potential source of information that could be used to improve the accuracy of risk ratings is the compliance reports that firms must prepare and maintain on-site under rules that became effective in 2004 (Compliance Program Rules), but do not have to file with SEC.[13]

These reports include information on the quality of the firms' compliance controls and any material weaknesses identified, which could be useful to OCIE for risk-rating purposes if OCIE were able to review these records regularly outside of routine examinations. Implementation of a prior recommendation to periodically obtain and review these compliance reports could potentially help OCIE better identify higher-risk firms.

Goal of Revised Examination Approach for Investment Companies and Investment Advisers Is to Identify and Shift Resources to Higher-Risk Firms and Compliance Issues

Following the detection of mutual fund trading abuses in the summer of 2003, OCIE revised its examination approach for investment companies and investment advisers. Specifically, OCIE shifted its examination approach from one that focused largely on the routine examination of all registered firms on an established schedule, regardless of risk, to one that targets resources on firms and issues that present the greatest risk of having compliance problems. Between 1998 and 2003, routine examinations accounted for about 90 percent of the approximately 10,400 investment company and investment adviser examinations OCIE conducted. During this period, OCIE generally tried to examine each firm at least once every 5 years.[14] However, the growth in the number of investment advisers, from 5,700 to about 7,700, and in the breadth of their operations did not allow OCIE to maintain this routine examination cycle. Also, OCIE concluded that routine examinations were not the best tool for broadly identifying emerging compliance problems, because firms were selected for examination based largely on the passage of time and not their particular risk characteristics.

To address these limitations, OCIE implemented a new risk-based examination approach in fiscal year 2005 that provides for more frequent routine examination of investment advisers determined to be higher-risk for compliance issues. Under this revised approach, OCIE's goal is to conduct at least one on-site, comprehensive, risk-based examination of all firms that have a higher-risk profile every 3 years. From those firms designated as lower-risk, OCIE randomly selects a sample each year to routinely examine. According to the 2007 "goals" memorandum—OCIE's key planning document for communicating examination priorities and guidance to examiners nationwide—OCIE targets more than three times the amount of examination resources to the routine examinations of higher-risk investment advisers (and their associated investment companies) than to the routine examination of lower-risk firms. Higher-risk firms represent about 10 percent of registered firms and 51 percent of assets under management. OCIE also now targets greater resources to sweep and cause examinations.

As part of its revised approach, OCIE began a pilot program in fiscal year 2006 that uses dedicated teams of two to four examiners to provide more continuous and in-depth oversight of the largest and most complex groups of affiliated investment companies and investment advisers. As of June 2006, OCIE officials said that a few select firms, representing approximately $1.5 trillion, or 4 percent, of assets under management in the United States, are currently participating in this voluntary program. Because these firms have been in the program for less than 12 months, we were unable to evaluate the effectiveness of OCIE's monitoring teams or this pilot. OCIE officials told us they plan on adding a limited number of additional firms and corresponding monitoring teams to the program by the end of 2007. Depending on the results of the pilot, the officials tentatively plan to have at least one and, perhaps, two monitoring teams in each field office.

To enhance OCIE's ability to identify and address emerging risks across the securities industry, in 2004 OCIE implemented a process intended to identify and map high-risk industry practices and compliance issues across the securities markets, including investment companies, investment advisers, and broker-dealers. SEC's Office of Risk Assessment (ORA) initially developed this process for agencywide use. In 2005, this process was automated, using a database application called Risk Assessment Database for Analysis and Reporting

(RADAR).[15] As used by OCIE, examiners in headquarters and regional offices identified and prioritized various risks to investors and registrants. OCIE staff then used RADAR to identify the highest-risk areas designated by examiners and then develop and recommend regulatory responses to address these higher-risk areas.[16] For example, OCIE officials said that they are addressing some risks by conducting examinations on the related issues and other risks by recommending that Market Regulation and Investment Management provide new rules or interpret existing ones. In addition, as part of OCIE's fiscal year 2007 goals memorandum, OCIE included information on the key risks identified through RADAR for each registrant type. OCIE examiners were directed to consider these risk areas as they plan examinations. OCIE officials said that they have not yet formally evaluated the effectiveness of RADAR for identifying new or resurgent compliance risks, as they have been largely focused on developing RADAR and the risk-assessment process itself. However, they said that the implementation of their recommended regulatory responses, through their own examination program and by other divisions and offices, would allow them to validate the risks identified through RADAR. OCIE officials said that they are considering developing a task force whose role, in part, would be to track the outcome of what OCIE recommends for risks entered in RADAR.

OCIE Is Taking Steps to Refine Its Method for Assessing the Compliance Risk Level of Investment Advisers but Faces Challenges

Accurately identifying compliance risk among registered investment advisers is critical to OCIE's revised approach to examinations, particularly for routine examinations. Because only a small number of low-risk firms are selected for routine examinations in a year, improperly categorizing investment advisers as lower-risk could lead to harmful practices' not being detected on a timely basis. To determine which firms are higher-risk and thus a priority for routine examination, every year OCIE queries its examination database and identifies those investment advisers that have been examined during the past 3 years and assigned a compliance risk rating of "high," indicating that their compliance controls have been assessed as "weak." These firms are automatically placed on the high-risk list and scheduled for routine examination within a 3-year period. However, because OCIE had only begun assigning risk ratings to firms in 2002 when it started using its risk-scorecard approach to evaluate compliance risks at individual firms, it was unable to assign risk ratings to all firms prior to revising the approach to routine examinations in 2004. Approximately 70 percent of registered investment advisers had not yet received a compliance risk-rating through a routine examination before OCIE implemented its new approach. Further, according to OCIE, its staff have not yet examined most of the more than 4,500 new investment advisers that have registered with SEC since fiscal year 2004.

To assign a risk rating for investment advisers that have never been examined by OCIE, OCIE uses an algorithm to calculate a numeric "score" for each firm based on certain affiliations, business activities, compensation arrangements, and other disclosure items that pose conflicts of interest.[17] Examples include participation or interest in client transactions, managing portfolios for individuals, and receiving performance fees. OCIE determines the risk profile of all registered investment advisers every year using the risk algorithm. Those that are designated as higher-risk through this method are added to the high-risk list and scheduled for routine examination within the next 3 years. At the start of fiscal year 2006, OCIE officials said they had identified about 10 percent of registered investment advisers as higher-risk. Slightly more than half of these were firms that had been routinely examined by OCIE within the last 3 years and given a risk-rating of "high" and slightly less than half were rated as higher-risk through the risk algorithm. A small percentage were firms OCIE had classified as higher-risk because of their large size. OCIE automatically designates the top 20 investment advisers according to assets under management as higher-risk.[18] According to OCIE officials, these larger firms are a priority because of the number of investors who could suffer adverse consequences as a result of any compliance problems at these firms.

Although the risk algorithm allows OCIE to determine an investment adviser's relative risk profile in the absence of a compliance risk rating, it is potentially limited because it does not measure the effectiveness of the investment adviser's compliance controls that are designed to mitigate conflicts of interest or other risks that could harm mutual fund shareholders. Rather, it relies on information that serves largely as proxy measures of the firm's compliance-related controls. OCIE has recognized these limitations and has taken some steps to evaluate the effectiveness of this methodology. OCIE officials told us they evaluate the accuracy of the risk ratings generated by the risk algorithm by comparing the results of completed routine examinations of firms initially presumed to be low-risk against the examination's outcome. According to data generated by OCIE, 91 percent of investment advisers that were initially rated lower-risk and examined in fiscal year 2006 retained the lower-risk designation after examination. OCIE officials said that they are reviewing the remaining 9 percent of examinations where the risk rating changed from lower to higher to determine the reasons for the change and whether they can use that information to improve the accuracy of the risk algorithm.

OCIE data also showed that 25 percent of all investment advisers that were initially rated higher-risk and examined during fiscal year 2006 retained their higher-risk rating, while the remaining 75 percent were reclassified as lower-risk. According to OCIE officials, one reason that the accuracy rate for predicting higher-risk firms appears low is that many of the firms on the higher-risk list, as previously discussed, were classified as higher-risk as a result of a prior examination. These firms likely took steps in the interim to improve their compliance controls, so OCIE officials expected that these firms would be rated as lower-risk after reexamination. However, the officials said that there are also many firms that had ratings assigned through the risk algorithm, and the fact that their ratings were changed from higher- to lower-risk after the examination demonstrates the limitations of the algorithm—it can determine which firms are at higher risk for compliance problems, but does not indicate the effectiveness of the firms' policies or procedures for mitigating these risks.[19]

To improve the accuracy of the risk-algorithm, OCIE initiated a sweep examination during 2007 of a sample of recently registered investment advisers that were identified as lower-risk and that had not yet been subject to a routine examination by OCIE. These reviews are typically targeted, 1-day reviews that allow examiners to obtain an initial assessment of these recently registered investment advisers' conflicts of interest, the related compliance policies and procedures these advisors use to manage these risks, and the capabilities of the firms' compliance and other personnel. OCIE anticipates that these limited-scope visits will assist examiners in determining whether a recently registered investment adviser's risk rating is accurate, and if it is not, will allow them to assign a more accurate risk rating and potentially identify additional information to refine the risk algorithm. According to OCIE officials, thus far, examiners have concluded over 225 of these reviews, with 85 percent of these resulting in firms' remaining classified as lower-risk and 15 percent being reclassified as higher risk and placed on a 3-year examination cycle. The officials said that they plan to make these sweep examinations a regular component of the examination program.

In addition, OCIE officials said that they are exploring ways to obtain and use additional sources of information that will allow them to further identify higher-risk firms.[20] OCIE officials told us they have purchased access to several commercial databases containing information on various data points, such as the performance of investment advisers, that OCIE does not otherwise have. OCIE officials said that they are currently assessing the usability of these databases for surveillance purposes, primarily to identify higher-risk firms. For example, if a firm's reported performance is significantly higher or lower than its peers, the officials said that performance could indicate that the firm's business processes deviate from the norm and require follow-up. Further, OCIE officials said that several OCIE and Office of Information Technology (OIT) staff are working on a project to identify other possible information sources that OCIE could use to better monitor investment companies and investment advisers. OCIE officials said that they are formalizing this effort by creating a Branch of Surveillance and Reporting, which will have staff permanently dedicated to the review and analysis of internal and external data sources to identify compliance risks at registered investment companies and investment advisers.[21]

The accurate prediction of each investment adviser's risk-level is critical to the protection of investors under the revised approach, as some firms rated lower-risk may never be routinely examined within a reasonable period of time, if at all, because of the sampling approach being used.[22] According to OCIE's review of 2006 examination data, 9 percent of investment advisers currently classified as lower-risk firms are actually higher-risk firms that should be scheduled to be examined within the next 3 years. Among newly registered advisers, the results of OCIE's targeted 1-day reviews show that the percentage of firms inappropriately characterized as lower-risk appears to be higher. OCIE's efforts to improve the capacity of the algorithm and obtain alternative sources of surveillance information could increase the likelihood that higher-risk firms will be identified and examined. However, neither the risk algorithm nor the alternative information sources OCIE is currently considering give OCIE any insights into the effectiveness of a firm's internal controls for mitigating identified compliance risks.

One potential source of information that might allow OCIE to assess the effectiveness of firms' internal controls is the reports registered investment companies and investment advisers are required to prepare at least annually under the Compliance Program Rules.[23] These rules require firms subject to the rule to adopt written compliance policies and procedures and review, at least annually, the adequacy of such compliance controls, policies, and procedures and the effectiveness of their implementation. Registered investment companies must designate a chief compliance officer responsible for giving the firm's board of directors a written report that, among other requirements, addresses the operation of the compliance controls of the investment companies and the controls of each of its service providers, including its investment adviser, and each material compliance matter that has occurred during the reporting period. Under the Compliance Program Rules, each investment company and investment adviser is required to maintain as part of its books and records any records documenting the firm's annual review of its compliance controls. OCIE staff currently review these compliance reports as part of the examination-planning process to learn about compliance issues identified by the firms and determine whether the firms have implemented corrective action. Currently, the rule does not require firms to submit the annual reports to the agency for its ongoing review. We previously recommended that SEC obtain and review these reports or the material weaknesses identified in them periodically rather than solely in connection with a planned examination.[24] OCIE officials noted that obtaining these reports on a regular basis would require rule making by SEC. We continue to believe, however, that using these reports outside of the examination process could potentially allow OCIE to improve its ability to identify higher-risk firms.

WITH SOME EXCEPTIONS, OCIE GENERALLY APPLIED EXIT PROCEDURES FOR THE PERIOD WE REVIEWED

Our review of investment company, investment adviser, and broker-dealer examinations completed during fiscal year 2003 through fiscal year 2006 found that examiners generally applied OCIE's exit procedures. OCIE's guidance on exit procedures gives examiners flexibility for communicating deficiencies and outcomes of examinations to registrants. These procedures include an exit interview, in which examiners inform registrants of deficiencies before closing an examination, and a closure notification letter, which communicates the outcome of the examination. Under certain circumstances, these procedures may not apply, such as when examiners refer their findings to Enforcement and are asked to forgo any or all of the procedures. OCIE management also has directed examiners to deviate from exit procedures under exigent circumstances, most recently for certain sweep examinations conducted during fiscal years 2003 through 2004 that addressed market timing and other newly emergent, high-risk compliance issues. OCIE officials told us that they did not inform the industry of their decision to forgo exit procedures for many of these sweep examinations, a situation that various industry participants told us confused the firms because they did not receive information on the status or outcome of the examination. We reviewed a sample of investment adviser and investment company examinations and a sample of broker-dealer examinations completed during fiscal year 2003 through fiscal year 2006. Based on this review, we estimated that examiners generally applied exit procedures. The exceptions were an estimated 9 percent of investment company and investment adviser examinations that were part of the market-timing and other related sweep examinations, as well as an estimated 7 percent of broker-dealer examinations where examiners either did not conduct these exit procedures or they did not provide evidence that they conducted them.

OCIE Guidance on Exit Procedures Gives Flexibility to Examiners for Communicating Examination Findings

OCIE has instituted specific exit procedures that give flexibility to examiners for communicating deficiencies and notifying registrants of the outcome of examinations. Prior to closing an examination, the guidance generally requires examiners to offer registrants an exit interview to inform them of any deficiencies that examiners found. According to a December 2001 memorandum, which formalizes OCIE's guidance for conducting exit interviews, these interviews are to ensure that registrants are informed of examiners' concerns at the earliest possible time, give registrants an opportunity to provide additional relevant information, and elicit early remedial action.[25] According to the guidance, OCIE's goal is to ensure that examiners inform registrants of all deficiencies prior to sending written notification of the examination findings, while at the same time giving examiners flexibility as to when to communicate their concerns. If examiners find deficiencies, they can communicate them either informally during the course of the fieldwork while on-site at the registrant, during a formal exit interview at the end of the on-site visit, or in an exit conference call after they complete additional analysis off-site. The guidance directs examiners to document these discussions in the examination's work papers and in the final examination report. OCIE's guidance for exit interviews also permits examiners to take into consideration the extent and severity of matters found during examinations when determining whether to conduct an exit interview. For example, when examiners refer a firm to Enforcement for securities law violations, in some cases Enforcement staff will ask the examiners to refrain from further discussions with the firm to protect the integrity of the impending investigation.

OCIE officials also told us that if the examiners did not identify any deficiencies to bring to the firm's attention once the on-site visit and subsequent fieldwork were complete, they were not expected to conduct formal exit interviews. Instead, examiners would let the firm know at the end of the on-site visit that they had not found any problems to date. If after completion of the off-site analysis, the examiners still did not identify any deficiencies, the guidance directs examiners to inform the firm of that fact prior to closing the examination.

Examiners formally close an examination by sending a "closure notification" letter to the firm. A closure notification letter can be a no further action letter, which indicates the examination concluded without any findings, or a deficiency letter, which cites any problems found. While the examiners may issue a deficiency letter and also refer some or all of the examination findings to Enforcement, in some cases, as with exit interviews, Enforcement staff may ask the examiners to refrain from sending a deficiency letter or exclude certain findings from a deficiency letter.

OCIE Deviated from Exit Procedures for Certain Market-Timing and Other Related Sweep Examinations, but Did Not Inform the Industry

OCIE officials said that they only direct examiners to deviate from established exit procedures when they believe it is in the best interest of the examination program and under exigent circumstances, such as during the period OCIE conducted sweep examinations of hundreds of firms to gather information on market timing and other newly emergent, high-risk compliance issues. OCIE officials said that in consultation with Enforcement staff, they decided for several of the market-timing and certain other concurrent sweep examinations to direct examiners not to conduct exit interviews or send closure notification letters. OCIE officials told us that they conducted these sweep examinations largely over fiscal years 2003 and 2004. As discussed later in this section, examination data we reviewed showed that some of these examinations were not completed until fiscal year 2005 or fiscal year 2006.

OCIE officials discussed the factors that contributed to their decision. First, they said that OCIE staff and examiners in the regional offices had little prior experience planning, conducting, and reporting on sweep examinations of such large scale and on such complex issues as market timing. At that time, OCIE did not have formal protocols in place to guide examiners when conducting sweep examinations. Second, the officials said that these sweep examinations involved a prolonged production of documents, data, and e-mails by firms and analysis of this information by OCIE and other SEC divisions and offices over periods as long as a year or more. For example, OCIE staff said that the review of initial documents provided by many firms often did not reveal any deficiencies, but the review of more detailed data a few months later did reveal deficiencies. OCIE officials said that if they had conducted an exit interview or sent a no further action letter based on the initial review of data, registrants would have stopped sending documents to the examination staff. As a result, the examiners would not have been able to detect the deficiencies that such information would have revealed. Third, OCIE officials said that to expedite the process for some groups of firms, they directed examiners not to write individual examination reports, which would have formed the basis for exit interviews and deficiency or no further action letters. Rather, they asked examiners to write a global report summarizing their findings.

Further, OCIE officials said that they did not inform the individual firms targeted during these sweep examinations or the industry generally of their decision to direct examiners to forgo exit procedures. We obtained the views of various industry participants representing investment companies, investment advisers, and broker-dealers on OCIE's decision. Several registrants said that the lack of communication during these sweep examinations was problematic and unsettling, as often they were unsure of the status of the examination, if they should be concerned about what OCIE was finding, or when they could assume the examination was over. Other industry representatives echoed these concerns and said that for any OCIE examination, early and ongoing communication with the examiners regarding any deficiencies identified, in addition to holding prompt exit interviews, is essential for the examination process to be effective and efficient. First, the representatives said that firms want to know immediately whether the examiners have identified any deficiencies so they can begin to address them as soon as possible. Second, if examiners identify deficiencies early, it allows the firm the opportunity to clarify any potential misinterpretations by examiners of the firm's policies, procedures, and practices before a deficiency letter is sent.

In the wake of the market-timing and other related sweep examinations, OCIE officials said they expect examiners to follow standard exit procedures for all sweep examinations, i.e., to conduct exit interviews to discuss any deficiencies, send deficiency or no further action letters, and make referrals to Enforcement as appropriate. In March 2006, OCIE issued formal guidelines for initiating, conducting, and concluding these examinations. As part of the guidelines, OCIE clarified that it expected sweep examinations to follow the same procedures as for other types of examinations. OCIE officials said that these expectations were reinforced with the issuance of an updated examination brochure (described in more detail below) in July 2006, which examiners are to provide registrants when beginning any examination and which details the exit procedures.

Examiners Generally Applied Exit Procedures during Review Period, with Some Exceptions Noted

Based on our review of a sample of investment company and investment adviser examinations and a sample of broker-dealer examinations completed during fiscal year 2003 through fiscal year 2006, we estimated that examiners generally applied exit procedures, with some exceptions. The exceptions were an estimated 9 percent of investment company and investment adviser examinations that were part of the market-timing and other related sweep examinations, as well as an estimated 7 percent of broker-dealer examinations where examiners either did not conduct these exit procedures or they did not provide evidence that they conducted them. In conducting this analysis, we analyzed examination data from two random samples of 129 examinations each, drawn from (1) the population of 8,107 investment company and investment adviser examinations and (2) the population of 3,044 broker-dealer examinations completed during fiscal year 2003 through fiscal year 2006. These samples allowed us to project the results of our review to the population of all investment company and investment adviser examinations and to the population of all broker-dealer examinations completed during this period at a 95 percent level of confidence.

We estimated that examiners held exit interviews to discuss deficiencies found in 79 percent of the investment company and investment adviser examinations completed during the review period (see figure 1). In addition, examiners did not conduct these interviews for reasons allowed under OCIE's exit interview guidance in an estimated 12 percent of the examinations. For example, in some cases examiners did not find any deficiencies during the examination and so were not required to conduct an exit interview. Instead, they were only required to inform the firm that they did not find any deficiencies and later send a no further action letter closing the examination.[26] Other reasons for not conducting exit interviews included referrals to Enforcement, where Enforcement staff directed the examiners to forgo the interviews, and other circumstantial reasons.

We estimated that for the remaining 9 percent of examinations, examiners did not conduct exit interviews because these examinations were part of the market-timing and other related sweep examinations where OCIE directed examiners to forgo these interviews, even though deficiencies were found.

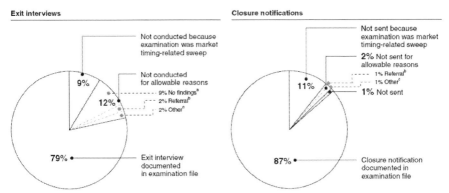

Source: GAO analysis of SEC examination data.

Note: Percentages may not add exactly because of rounding. All estimated percentages in this table have margins of error of plus or minus 8 percent or less.

[a]Examiners were not required to conduct a formal exit interview because they did not find any deficiencies during the examination.

[b]Examiners referred deficiencies found to Enforcement, whose staff requested that examiners forgo conducting the exit interview, sending a deficiency letter, or both.

[c]Other reasons include more circumstantial reasons OCIE examiners did not conduct exit interviews or send closure notifications. For example, in one case, the firm did not produce requested documents in a timely manner during a sweep examination conducted by headquarters staff, and the examination team closed the exam and referred the firm to a regional office, which began a new examination. We did not request additional information on the new examination.

Figure 1. Estimated Percentage of Investment Company and Investment Adviser Examinations Where Examiners Conducted Exit Interviews and Sent Closure Notifications, Fiscal Years 2003 to 2006.

We also analyzed the frequency with which examiners sent closure notifications to investment companies and investment advisers and estimated that examiners sent either a deficiency or a no further action letter in 87 percent of the examinations completed during the review period. The predominant reason for the higher rate of closure notifications sent compared with exit interviews conducted was that examiners sent no further action letter to firms when no deficiencies were found.

Examiners did not send closure notification letters in an estimated 11 percent of examinations (14 of the 129 examinations we reviewed) because the examinations were part of the market-timing and other related sweep examinations and OCIE had directed examiners not to send any letters. We found that of these 14 examinations, 11 concluded in fiscal year 2004, 2 concluded in fiscal year 2005, and 1 concluded in fiscal year 2006. Examiners did not send closure notifications for allowable reasons in an estimated 2 percent of examinations, and in an estimated 1 percent, we found no evidence that closure notifications were sent and no legitimate reason why they should not have been sent.

For the population of broker-dealer examinations, we estimated that examiners conducted exit interviews and sent closure notifications in an estimated 82 percent and 88 percent, respectively, of examinations conducted during the review period (see figure 2). Examiners did not conduct exit interviews or send closure notifications for reasons allowable under OCIE guidance, such as related to referrals to Enforcement, in an estimated 11 percent and 9 percent, respectively, of the examinations.

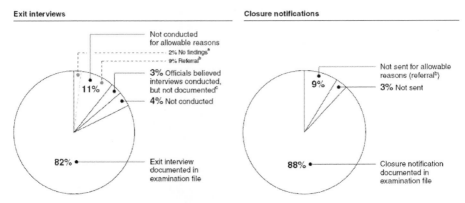

Source: GAO analysis of SEC examination data.

Note: Percentages may not add exactly because of rounding. All estimated percentages in this table have margins of error of plus or minus 8 percent or less.

[a]Examiners were not required to conduct a formal exit interview because they did not find any deficiencies during the examination.

[b]Examiners referred deficiencies found to Enforcement, whose staff requested that examiners forgo conducting the exit interview, sending a deficiency letter, or both.

[c]"OCIE officials believed interviews conducted, but not documented," refers to those cases where OCIE officials believed that examiners conducted exit interviews, but were not able to provide documentation of them.

Figure 2. Estimated Percentage of Broker-Dealer Examinations Where Examiners Conducted Exit Interviews and Sent Closure Notifications, Fiscal Years 2003 to 2006.

In an estimated 7 percent of examinations, we did not find evidence of an exit interview when OCIE guidance indicated one was warranted. However, this estimate includes the 3 percent of cases where OCIE officials told us they believed examiners conducted the interviews but did not document the discussion.

For the other estimated 4 percent, we found no evidence that exit interviews were held and no legitimate reason why they should not have been held. In addition, we found no evidence that closure notifications were sent in an estimated 3 percent of examinations and no legitimate reason why they should not have been sent.

OCIE IMPLEMENTED NEW INITIATIVES INTENDED TO IMPROVE COMMUNICATION AND COORDINATION

OCIE has implemented several initiatives since January 2006 designed to improve internal and interagency coordination and communication with registrants. For instance, OCIE has undertaken efforts that include developing tools and protocols to avoid duplication of examinations and forming interdivisional committees intended to improve referrals to Enforcement. Other initiatives include establishing a "hotline" for registrants and formalizing a new requirement to notify registrants when an examination extends past 120 days. Our review indicated that examiners generally complied with this new notification requirement. Some industry participants who provided their views on OCIE's initiatives expressed hesitations about using the new hotline. Specifically, several participants questioned the independence of the new hotline, because it is located within OCIE's Office of the Chief Counsel and not in another SEC office or division.

OCIE Has Recently Implemented Initiatives to Increase Communication with Registrants and Improve Interagency Coordination

In May 2006, the SEC Chairman testified before the House Financial Services Committee on several reforms designed to improve preexamination planning and increase the transparency of SEC's examination program. We found that OCIE has generally implemented the following reforms and additional protocols and tools that are intended to improve coordination across SEC headquarters and regional offices and with other key SEC divisions.

- To minimize the number of firms selected for multiple sweep examinations and to provide advance notice to the commission regarding planned sweep examinations, OCIE developed a formal review and approval process for sweep examinations that is detailed in the March 2006 sweep examination guidance previously discussed. As part of this guidance, OCIE field examiners and OCIE staff are directed to submit a list of firms to OCIE management that they propose to include as part of the sweep examinations. OCIE management is responsible for comparing the list of proposed firms against a master list of firms subject to ongoing or recently completed sweep examinations to ensure that that the firms are not bearing an undue share of examination focus, given the nature of their business and OCIE's risk assessment. Once OCIE has approved the proposed sweep examination and the targeted firm, the new guidance directs OCIE to provide the proposal to Market Regulation, Investment Management, or Enforcement for notification and to obtain comments. Finally, OCIE is to provide the Commission an information memorandum summarizing the proposed sweep examination and its objectives. The memorandum directs staff to allow the Chairman and Commissioners time to review the memorandum and ask questions before commencing the sweep examination. We reviewed four of these memorandums, which discussed the time frames for the sweep, the issues OCIE planned to investigate and the methodology it would use, and the firms it planned to include.

• SEC, NASD, and NYSE have developed a database, maintained by NASD, which collects data on examinations conducted by SEC, NYSE, and NASD on 170,000 broker-dealer branch offices. OCIE officials said that examiners are now using this database when planning examinations to avoid dual examinations of the same branch office (with the exception of the broker-dealers selected for review as part of SEC's oversight program of the SROs). Further, as part of the pilot program for assigning permanent monitoring teams to the largest investment company and investment adviser complexes, each firm's monitoring team is responsible for conducting all examinations related to the firm, including examinations at branch offices in different areas of the country. OCIE officials said that prior to the implementation of this program, examiners from different regional offices would conduct separate examinations of the firm's branch offices, which resulted in duplication and imposed a burden on the firm.

• To improve coordination with other key SEC divisions, OCIE officials said they have designed a new training program for fiscal year 2007 that is designed to educate examiners about rules affecting investment companies, investment advisers, and broker-dealers. The four scheduled courses are taught by Investment Management and Market Regulation staff and focus on rules that are new or about which examiners have frequent questions in the course of conducting examinations for compliance with these rules. Second, OCIE and Enforcement have established interdivisional committees in headquarters and the regional offices in late 2006 and 2007 to bring more transparency and consistency to the decisions made to pursue OCIE referrals to Enforcement about investment companies, investment advisers, and broker-dealers. According to joint guidance issued by OCIE and Enforcement in November 2006, the responsibilities of these committees include discussing new referrals to understand their strengths and weaknesses and reviewing those examinations referred to Enforcement that have not resulted in an investigation or enforcement action. In headquarters, these committees also include staff from Investment Regulation and Market Regulation, whose role is to provide insight with respect to referrals that involve novel fact patterns or applications of the law. OCIE has also taken the following measures that are intended to improve communication with registrants.

- In January 2006, OCIE established an examination hotline where registrants can call or e-mail anonymously to ask questions about their specific examinations or other issues, lodge complaints, or make comments. To preserve anonymity of the registrants, OCIE does not keep a formal log of calls and e-mails to share with OCIE management although staff take notes on the calls. OCIE officials told us that examples of complaints and concerns to date have included duplicative requests, complaints about examiners, and questions about public statements made by OCIE officials about the examination program. We discuss registrants' views of the effectiveness of the new hotline later in this section.

- In July 2006, OCIE began requiring examiners to contact registrants when examinations extend beyond 120 days to discuss the status of the examination, the likely schedule for completion, and the date of the final exit interview. Previously, OCIE officials told us that examiners had no notification requirement but as a best practice, tried to contact the registrant if the examination extended beyond the usual 90 days it took to complete most examinations. However, because of the increasing complexity of firms and the increased emphasis on sweep examinations, both of which require additional analysis on the part of examiners and can increase the time needed to complete an examination, OCIE decided to formalize this practice so that firms would be fully aware of the status of the examination. We discuss the extent to which examiners complied with the new notification requirement later in this section.

- OCIE officials said that they have made more information publicly available on the examination program and current compliance issues. In July 2006, OCIE issued a revised examination brochure, which provides more detailed information to registrants about the examination process, including the 120-day notification procedure and exit procedures. Later, in January 2007, SEC issued a guide, prepared by OCIE, to assist broker-dealers in their efforts to comply with anti-money-laundering laws and rules. Finally, in June 2007, OCIE issued its first *Compliance Alert* letter to chief compliance officers (CCO) of investment companies and investment advisers as part of its CCO outreach program.[27] These letters, which OCIE officials said they plan to issue twice a year, are intended to describe areas of recent examination focus and certain issues found during investment company, investment adviser, and broker-dealer examinations.

Examiners Generally Followed OCIE's New Procedure to Notify in 120 Days Where Applicable

We reviewed OCIE examination data to determine the extent to which examiners followed the new 120-day notification procedure for investment company, investment adviser, and broker-dealer examinations and determined that examiners generally followed this procedure where applicable since its implementation on July 31, 2006. As previously discussed, OCIE implemented the new 120-day notification procedure to better inform registrants of the status of examinations that would not be completed within 120 days. OCIE has directed examiners to contact the firm on or about the 120th day after the completion of the on-site visit to discuss the status of the examination and the likely schedule for completing the examination and conducting an exit interview. OCIE officials said that this procedure largely was intended to address those instances when an examiner left the firm after the on-site portion of the examination and did not have further contact with the firm while conducting subsequent analysis. OCIE examiners are instructed to call the firm in these cases to update the firm on the examination and should document the discussion in a note to the examination file. However, OCIE officials said that sometimes an examination will extend beyond 120 days because OCIE is waiting for the firm to produce documents or data. In those cases, the firm knows that the examination is still ongoing and examiners are not expected to call on or about the 120th day.

We identified a total of 13 closed examinations that had lasted 120 days or more in the period between the date OCIE implemented the new procedure (July 31, 2006) and the date OCIE provided us its records (February 2, 2007). These 13 cases included 10 investment company and investment adviser examinations and 3 broker-dealer examinations. In 7 of the 13 examinations, examiners either contacted the firm on or around the 120th day of the examination or otherwise already had ongoing communication with the firm because they were waiting for documents or other data from the firm. In 5 of the 13 examinations, examiners did not provide 120-day notification for allowable reasons. For example, in two of these five cases, the examiners referred their findings to Enforcement staff, who asked the examiners to cease contact with the firm. In the other three of these five cases, further contact was not warranted, either because the firm decided to withdraw its registration or the examiners only reviewed available data about the firm in SEC's offices and never contacted the firm to open a formal examination. In the last of the 13 examinations, examiners did not contact the firm on or about the 120th day.

Comments from Industry Participants on OCIE's New Initiatives Revealed Concerns about the Independence of the New Hotline

We contacted various industry participants representing investment companies, investment advisers, and broker-dealers to gather their views on OCIE's recent initiatives. They generally expressed support for these initiatives, but some expressed hesitations about using the new hotline. More specifically, several registrants viewed the new hotline as a positive step by OCIE to provide an additional channel of communication, and at least one registrant reported finding the hotline very useful. However, others expressed concern about the independence of the staff that operate the hotline, because, as previously discussed, OCIE's hotline is staffed by attorneys in OCIE's Office of the Chief Counsel. Other industry participants questioned the utility of the hotline as a tool for addressing issues that are of concern to them. For example, they said that they often have concerns about the interpretation of SEC rules by examiners during examinations. They said that when these issues arise, they would like OCIE to consult with Market Regulation —the SEC division that writes and interprets the rules for broker-dealers—to ensure that examiners interpret these rules appropriately. However, they said that they do not perceive that this consultation currently occurs, and as a result, have doubts that calling the hotline would result in an effort by OCIE to obtain clarification. Further, they said that it is important to resolve concerns about rule interpretations while an examination is ongoing and obtain the views of Market Regulation early on, before the exit conference occurs or a deficiency letter is sent. Similarly, another group of registrants thought that OCIE was unresponsive to their past concerns and did not see the hotline as a valuable tool for addressing these concerns.

OCIE officials told us they decided to locate the hotline in their Office of the Chief Counsel because this office is the ethics office for OCIE. OCIE managers thought it was important to keep the hotline in a centralized location, as an issue could arise that involved any one of OCIE's examination programs (such as the programs for investment companies and investment advisers or broker dealers). In addition, according to OCIE officials, the Associate Director and Chief Counsel reports directly to the OCIE director, and therefore the Chief Counsel's Office can exercise a great deal of institutional autonomy when determining how to handle the calls or e-mails received. As discussed earlier, OCIE officials said the Chief Counsel's Office does not keep a formal log of contacts, to better preserve the anonymity of registrants. Finally, the officials said that the issues that registrants bring to them may be legally sensitive, so it made sense that the Chief Counsel's office evaluates them first to determine how to best address them.

In contrast to OCIE, NASD has created an Office of the Ombudsman to receive and address concerns and complaints, whether anonymous or not, from any source concerning the operations, enforcement, or other activities of NASD. The Office of the Ombudsman is an independent office within NASD that reports directly to the Board of Directors. As part of its responsibilities, the Office of the Ombudsman also provides summary information on the development of trends based on complaints, which may support resulting system change. By locating the hotline in an office or division that is independent of OCIE, OCIE could lessen registrants' concern about the independence of that staff who operate the hotline and thus encourage greater use of it. Besides assisting callers with any complaints, the independent office could periodically summarize information from complaints and concerns for OCIE, while preserving the anonymity of the contacts. Such information could allow OCIE management to identify and respond to any trends in this information and potentially improve the examination program.

CONCLUSIONS

In the aftermath of the widespread trading abuses that surfaced in the mutual fund industry in late 2003, OCIE has taken steps to make its approach to examining investment companies and investment advisers more risk-based. While such an approach may provide a basis for OCIE to allocate its limited resources to examine firms that are designated as higher risk for compliance problems, the effectiveness of the program largely depends on OCIE's ability to accurately determine the risk level of each investment adviser. Since many firms rated lower-risk are unlikely to undergo routine examinations within a reasonable period of time, if at all, harmful practices could go undetected if firms are inappropriately rated as lower-risk. The risk algorithm that OCIE employs to predict the level of risk for the majority of investment advisers is potentially limited in that it relies on proxy indicators of compliance risks without incorporating information about the relative strength of a firm's compliance controls, information that is critical to assessing a firm's risk level. OCIE has recognized this limitation and has started to take steps to enhance the effectiveness of the risk algorithm to accurately predict risk levels by seeking additional information that could improve OCIE's ability to identify higher-risk firms. Based on fiscal-year-2006 data, OCIE's internal assessment showed that the risk algorithm has a 91 percent accuracy rate for predicting lower-risk ratings but appears to have a lower accuracy rate when considering newly registered investment advisers. Further, the accuracy rate for higher risk firms was 25 percent—in part because the risk-rating did not incorporate information on the firms' ability to mitigate the compliance risks identified. The results of these initial analyses indicate that continued assessing and refining the risk algorithm is warranted. As we have previously recommended, one potential source OCIE might consider, as it continues to enhance its methods that assess risk, is the documentation associated with the compliance review that firms must conduct under the Compliance Program rules. We recognize that SEC would first have to require that firms submit these reports to SEC through rule making. However, we continue to believe that using them could potentially allow OCIE to improve its ability to identify higher-risk firms. As part of the revised examination approach, OCIE has also implemented a process intended to identify, map, and develop regulatory responses to high-risk industry practices and compliance issues across the securities markets, although it has not yet developed a formal approach to evaluate the effectiveness of this process for identifying new or resurgent compliance risks. Implementing the task force that OCIE is currently considering could facilitate such an assessment.

Our review found that OCIE examiners generally followed OCIE guidance for conducting exit procedures during the period reviewed, with a major exception for market-timing and other related sweep examinations conducted largely over fiscal years 2003 and 2004, with a few concluding in fiscal year 2005 and fiscal year 2006. OCIE directed examiners to forgo these exit procedures. OCIE guidance provides management and examiners flexibility in determining when and how to communicate deficiencies to registrants and is responsive to Enforcement's directives. However, by not providing the industry any notice or explanation of the decision to forgo these procedures for certain market-timing and other related sweep examinations, OCIE unnecessarily created concern and confusion for some registrants during this difficult time. Going forward, OCIE has directed its examiners to follow standard exit procedures for all sweep examinations.

Since January 2006, OCIE has generally implemented a number of initiatives to improve coordination and communication. Ongoing monitoring and reassessing of these initiatives by OCIE is important to ensure that they are achieving their intended objective. For example, OCIE's procedure to notify firms when examinations continue beyond 120 days could help mitigate the uncertainty firms told us they experience when examiners leave the firm and do not update the firm on the status of examinations for long periods of time. While our review of 13 examinations revealed general compliance with this notification procedure, OCIE must ensure that examiners continue to adhere to the requirement in the future. Another new initiative, the examination hotline, could give registrants an effective means to communicate concerns or complaints about the examination program, but several registrants reported reluctance to use it because the hotline was located in and staffed by OCIE. Their concerns about OCIE's receiving their complaints or concerns included a perceived lack of impartiality. Locating the hotline in a division or office that is independent of OCIE could encourage greater use and increase effectiveness. Further, this new office could analyze the contact information and give OCIE management with information summarizing trends generated from analysis of complaints or inquiries, information that OCIE could use to improve its examination programs.

APPENDIX I: SCOPE AND METHODOLOGY

To better understand the Office of Compliance Inspections and Examination's (OCIE) revisions to the examination approach for investment companies and investment advisers after 2003 and the process OCIE implemented to better identify compliance risks across the securities markets, we obtained and analyzed information from OCIE on its revised examination approach for investment companies and investment advisers and a new, examiner-driven process for identifying emergent or resurgent systemic risks to investors and SEC registrants. Specifically, we reviewed OCIE's internal planning documents, a memorandum from OCIE to the Commission, and data OCIE generated as part of an internal evaluation on its methodology for assessing compliance risk at investment advisers. We did not verify these data. We also observed a demonstration of the information technology application OCIE uses to conduct its annual risk-assessment process, reviewed prior GAO reports, and interviewed OCIE and other Securities and Exchange Commission (SEC) officials.

To identify OCIE's exit procedures and assess the frequency with which examiners have followed these procedures when conducting investment company, investment adviser, and broker-dealer examinations, we reviewed OCIE's guidance to examiners and interviewed officials from OCIE and industry participants representing investment companies, investment advisers, and broker-dealers, and analyzed examination data. We focused our analysis on these registrants because they comprise over 95 percent of all SEC registrants and OCIE expends most of its examination resources on these entities. We analyzed broker-dealer examinations separately from investment company and investment adviser examinations because the two examinations areas have different types of examinations and are managed separately within OCIE.

We obtained data from OCIE on all investment adviser, investment company, and broker-dealer examinations completed during fiscal year 2003 through fiscal year 2006. We chose to review this period because it included a time when OCIE did not require examiners to apply their usual exit procedures for certain sweep examinations as well as periods when OCIE said that it applied the exit procedures to most examinations. We selected random samples of 129 examinations each from the population of 8,107 investment company and investment adviser examinations and the population of 3,044 broker-dealer examinations. This sample size allowed us to project our results from these two samples to the two respective populations at the 95 percent level of confidence. All estimates have margins of error of plus or minus 8 percent or less. To determine the extent to which examiners conducted exit interviews and sent closure notifications in our samples, we reviewed the electronically available examination reports and, where necessary, asked OCIE for additional documentation from the examination files. The results of our analysis for each of these two registrant types are limited to estimates of this combined 4-year time frame. In conducting our analysis, we conducted a data reliability assessment of the data OCIE provided us and determined they were reliable for our purposes.

To identify OCIE's recent initiatives to increase communication with registrants and improve the examination program, including the frequency with which examiners have followed OCIE's new notification requirement for examinations that continue longer than 120 days, we reviewed documentation obtained from OCIE, including memorandums to the Commission, internal OCIE guidance, and the revised examination brochure. We also analyzed examination data related to the 120-day notification procedure and interviewed officials from OCIE and the industry participants previously discussed. In determining the frequency that examiners have complied with the new 120-day notification procedure, we obtained data from OCIE of all of the investment adviser, investment company, and broker-dealer examinations conducted between July 31, 2006, the day the policy was implemented, and February 2, 2007, the day OCIE gave us the records. We identified all closed examinations that lasted 120 days or more, thus triggering the 120-day notification requirement. Because there were only 13 closed examinations where the procedure was applicable, we reviewed all of them. We first reviewed the electronically available reports for evidence of the notification and, in those cases where we did not find evidence, asked OCIE for additional documentation from the examination files. In conducting our analysis, we conducted a data reliability assessment of the data OCIE provided us and determined they were reliable for our purposes.

We conducted our work in Washington, D.C., between September 2006 and July 2007 in accordance with generally accepted government auditing standards.

REFERENCES

[1]   SEC regulates investment companies and investment advisers under the Investment Company Act of 1940, the Investment Advisers Act of 1940, the Securities Act of 1933, and the Securities Exchange Act of 1934. The Investment Company Act and the Investment Advisers Act requires certain investment companies and investment advisers, respectively, to register with SEC and thus subject their activities to SEC regulation. Broker-dealers are required to register with SEC and are subject to SEC regulation under the Securities Exchange Act of 1934.

[2]   We discussed SEC's response to the surfacing of these widespread abuses previously. See, for example, GAO, *Mutual Fund Industry: SEC's Revised Examination Approach Offers Potential Benefits, but Significant*

Oversight Challenges Remain, GAO-05-415 (Washington, D.C.: Aug. 17, 2005) and GAO, *Mutual Fund Trading Abuses: Lessons Can Be Learned from SEC Not Having Detected Violations at an Earlier Stage*, GAO-05-313, (Washington, D.C., Apr. 20, 2005).

[3] Protecting Investors and Fostering Efficient Markets: A Review of the S.E.C. Agenda Before the H. Comm. on Financial Services, Statement of Christopher Cox, Chairman, Securities and Exchange Commission, 109th Cong. 45-46 (2006).

[4] Compliance risks refers to the propensity of an SEC registrant to be in violation of federal securities laws and regulations or, where applicable, the rules of a governing self-regulatory organization.

[5] OCIE assigns risk ratings to investment advisers, but not investment companies. Many investment companies have few employees and rely on investment advisers to perform key functions such as providing management and administrative services. When OCIE examines an investment adviser, it generally examines related investment companies concurrently. OCIE officials estimated that about one-third of registered investment advisers have received applicable risk ratings from an examination as of September 2006.

[6] Currently, the use of these reports is limited to the routine examinations of investment companies and investment advisers, where OCIE examiners review the reports as part of the examination planning process to learn about compliance issues identified by these firms. See GAO-05-313, p. 35, for previous discussion of these reports and our related recommendation.

[7] Other compliance issues that surfaced during this time included the late trading of fund shares and the misuse of material, nonpublic information.

[8] Percentages do not add exactly to 100 percent due to rounding.

[9] Prior to 2002, routine examinations typically focused on discrete areas that staff viewed as representing the highest risks of compliance problems that could harm investors.

[10] In July 2007, after the completion of our fieldwork, NASD and the member regulation, enforcement and arbitration functions of NYSE consolidated to become the Financial Industry Regulatory Authority.

[11] OCIE undertakes SRO inspections in order to evaluate whether an SRO is (1) adequately assessing risks and targeting its examinations to address those risks, (2) following its examination procedures and documenting its work, and (3) referring cases to enforcement authorities when appropriate.

[12]   OCIE also conducts surveillance examinations, which are generally broker-dealer oversight examinations that occur slightly more than 12 months after the examination.

[13]   Rule 38a-1 applies to registered investment companies, including business development companies. See 17 C.F.R. §§ 270.38a-1 (2006). Rule 206(4)-7 applies to registered investment advisers. See 17 C.F.R §275.206(4)-7. Prior to the adoption of these rules, investment advisers were already subject to requirements to maintain written compliance polices and procedures in certain areas. See *Compliance Programs for Investment Companies and Investment Advisers*, 68 *Federal Register* 74714, 74715 n. 14 (Dec. 24, 2003) (adopting release), for a list of such requirements.

[14]   During 2003, OCIE began to address these concerns by establishing a 2-, 4-, or 5-year examination cycle based on the size or risk level of the investment adviser. However, this cycle was not fully implemented before OCIE made significant changes to its examination program for investment companies and investment advisers as described in this section.

[15]   OCIE developed the RADAR application in conjunction with staff from the Office of Risk Assessment (ORA) and OIT. In 2005 and 2006, RADAR was a database application; in 2007, OCIE and OIT staff enhanced RADAR to make it a Web-based application.

[16]   OCIE officials said that the risks entered into RADAR by SEC examination staff and managers are based on information learned during examinations and constitute non-public information.

[17]   The risk algorithm, developed by OCIE and the Office of Economic Analysis, is a formula using values of various factors to derive a relative ranking for the firm's compliance risk.

[18]   Combined, these 20 investment advisers have $8.9 trillion in assets under management, about 28 percent of all registered investment advisers' assets under management.

[19]   OCIE officials said that the composition of the higher-risk risk firms examined reflected the composition of the total firms rated higher-risk at the start of fiscal year 2006, in that slightly more than half were firms that had received a higher-risk rating through routine examination and slightly less than half had received higher-risk rating through the risk algorithm.

[20]   In 2005, SEC considered the development of a surveillance program for OCIE to gather and analyze additional information from investment companies and investment advisers outside of the data collected in OCIE's usual examination and reporting process. However, OCIE officials told us that SEC decided to postpone this effort, which would have imposed

potentially significant costs for SEC and firms and required formal rule making to implement, in favor of obtaining access to third-party databases.

[21]   OCIE officials told us they are planning to staff the new Branch of Surveillance and Reporting with a branch chief and four analysts.

[22]   OCIE officials said that when preparing to generate the random sample of investment advisers rated as lower-risk, they first remove from the universe those firms that were selected and routinely examined the previous year.

[23]   17 C.F.R. §§ 270.38a-1 and 275.206(4)-7 (2006).

[24]   GAO-05-313, 35.

[25]   Prior to the December 2001 memorandum, examiners were not required to offer an exit interview.

[26]   Several of the examinations in our sample which resulted in no deficiencies were market-timing and other related sweep examinations where OCIE officials told us that in many cases examiners did not provide any indication of the outcome of the examination regardless of whether any deficiencies were found.

[27]   SEC implemented the CCO Outreach program in 2005. The program is jointly sponsored by OCIE and Investment Management and is designed to enable the Commission and its staff to better communicate and coordinate with the CCOs of investment companies and investment advisers.

INDEX

E

F

G

H

I

O

P

T